The Finite Element Method
A Basic Introduction

The Finite Element Method
A Basic Introduction

K. C. Rockey, MScEng, PhD, CEng, FICE, FIStructE, MIMechE
Professor and Head of Department of Civil and Structural Engineering,
University College, Cardiff

H. R. Evans, MSc, PhD
Department of Civil and Structural Engineering,
University College, Cardiff

D. W. Griffiths, BScTech, MSc, AMCST, CEng, MIMechE
Department of Civil and Structural Engineering,
University College, Cardiff

D. A. Nethercot, BSc, PhD
Department of Civil and Structural Engineering,
University of Sheffield

A HALSTED PRESS BOOK

JOHN WILEY & SONS
NEW YORK

Granada Publishing Limited
First published in Great Britain 1975 by Crosby Lockwood Staples
Frogmore St Albans Herts
and 3 Upper James Street London W1R 4BP

Published in the U.S.A. by
Halsted Press, a division of
John Wiley & Sons Inc., New York

Printed in Great Britain

Library of Congress Cataloging in Publication Data
Main entry under title:

The Finite Element Method.

  'A Halsted Press Book.'
  Bibliography: p.
  1. Structures, Theory of.   2. Finite element
method.   I. Rockey, Kenneth Charles.
TA646.F5 1974   624'.171   74–6671
ISBN 0–470–72927–9

Preface

In recent years, the finite element method has become widely accepted by the engineering professions as an extremely valuable method of analysis. Its application has enabled satisfactory solutions to be obtained for many problems which had hitherto been regarded as insoluble, and the amount of research effort currently being devoted to the finite element method ensures a rapidly widening field of application.

Unfortunately, and perhaps to a certain extent because of the high level of research activity, a certain mystique surrounds the finite element method at the moment. The object of this book is to remove this mystique, so that engineers, either by developing their own computer programs or by the intelligent application of some of the excellent programs already available, will utilise the method to its full potential. The book is directed to those people who have no experience at all of the finite element method and those who, while having used some of the existing programs to obtain solutions in specific cases, are unfamiliar with the theoretical basis of the method.

The book is based on course notes prepared by the authors for a week-long teaching course at University College, Cardiff. This course has been run successfully four times and has been very well attended by people from a variety of engineering backgrounds, both from this country and from overseas. The authors are greatly indebted to the engineers who have attended these courses for the interest that they have shown; the many valuable suggestions made during the lively discussion periods have been fully implemented in preparing this text.

The aim throughout has been simplicity of presentation. The first three chapters have been devoted entirely to matrix analysis of skeletal structures, since a sound knowledge of matrix techniques is essential for an understanding of finite elements. Chapter 4 discusses the change in philosophy between the matrix analysis of skeletal structures and the finite element analysis of continua, where a natural sub-division of the structure no longer exists. In the following chapters, i.e. Chapters 5 to 9, the stiffness matrices for a number of finite elements are developed and presented, and these elements will enable the reader to solve a wide range of problems; Chapter 10 discusses how such solutions can be carried out efficiently on a digital computer. Finally, in Chapters 11 and 12 a number of more advanced topics are discussed, the aim being to give the reader a comprehensive picture of the current state of the art in the finite element field; the quoted references will enable the reader to obtain more details of any topic that is of particular interest to him. It is hoped that the book will stimulate the reader's interest in the method sufficiently for him to read some of the excellent research papers frequently published on various applications and developments currently being investigated in the finite element field.

The book gives little guidance on what type of sub-division and how many elements should be employed in any particular case. This omission is deliberate because most analyses, particularly in Civil Engineering, are of one-off structures, and there is extreme danger in generalising on the number of elements used in any particular case. The authors would propose that, faced with the choice of mesh for any particular problem, three or four different meshes should be used, each of different fineness, so that the rate of convergence of the results can be established and some confidence obtained in the sub-division finally employed.

In the finite element method a modified structural system consisting of discrete (finite) elements is substituted for the actual continuum and thus the approximation is of a physical nature. There need be no approximation in the mathematical analysis of this substitute system. By contrast, in the finite difference method the exact equations of the actual physical system are solved by approximate mathematical procedures.

Finally the authors wish to thank Mrs M. W. Ellis who typed the manuscript and Mr D. C. Jones, Mrs M. Dover and Mrs N. Woodward who prepared the drawings.

K.C.R.
H.R.E.
D.W.G.
D.A.N.

July 1974

Contents

Notation Index

A	area
$[A]$	coefficient matrix associated with displacement function, see Appendix 1
$[B]$	matrix relating element strains to element nodal displacements, see Appendix 1
$[D]$	elasticity matrix, see Appendix 1
D_x, D_y, D_{xy}, D_1	bending rigidities of plate
E	Young's modulus
F	force
$\{F\}$	vector of nodal forces
F_x, F_y, F_z	forces in x, y and z directions
G	shear modulus
$[H]$	stress–displacement matrix
I	second moment of area
k	stiffness
k_{ij}	term in $[K]$ located in row i and column j
$[K]$	stiffness matrix
L	length
L_1, L_2, L_3	area co-ordinates, see Chapter 11

M_x, M_y, M_{xy}	internal bending moments
N_x, N_y	shape functions, see Chapter 11
t	thickness
$[T]$	transformation matrix
T_x, T_y, T_z	moments about x, y and z axes
u, v, w	displacements along x, y and z axes
x, y, z	rectangular Cartesian co-ordinate system
α_1, α_2	constants used in displacement function
$\gamma_{xy}, \gamma_{yz}, \gamma_{zx}$	shear strains
δ	displacement
$\{\delta\}$	vector of nodal displacements
Δ	area of element
ε	strain
$\varepsilon_x, \varepsilon_y, \varepsilon_z$	direct strains
$\theta_x, \theta_y, \theta_z$	rotations about x, y and z axes
$\phi, \phi_{x\bar{x}}, \phi_{x\bar{y}}$	angles between local and global axes systems
σ	stress
$\sigma_x, \sigma_y, \sigma_z$	direct stresses
$\tau_{xy}, \tau_{yz}, \tau_{zx}$	shear stress
v	Poisson's ratio
$[\ \ ]$	indicates a matrix
$\{\ \ \}$	indicates a one-dimensional array, row or column vector
$[^e], \{^e\}$	matrix, vector relating to a single element
$\bar{x}, \bar{u}$	global quantities
$[\ \ ]^T$	transpose of matrix
$\{\ _i\}, x_i$	indicates quantities associated with node i
(x, y)	indicates quantities are functions of x and y
$[\ \ ]^{-1}$	inverse of matrix
w^{rb}	relates to rigid body component of displacement, see Chapter 11
$w^{ss}, [B]^{ss}$	relates to simply supported component of displacement, see Chapter 11

CHAPTER ONE

Basic Concepts

Introduction

Many engineering structures are composed of a series of individual members which are connected together at a number of points. Such structures are called 'skeletal' structures, the points at which the individual members are connected being referred to as 'node points'. Examples of such structures are the continuous beam and the multi-storey frame shown in Figure 1.1. Engineers have long appreciated

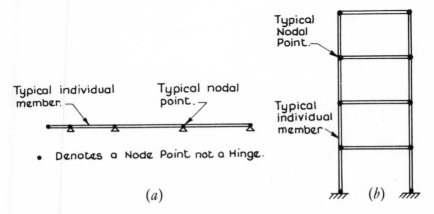

Fig. 1.1. Typical skeletal structures. *a* Continuous beam. *b* Multistorey frame

that the analysis of these skeletal structures can be carried out by first considering the behaviour of each individual element independently and by then assembling the elements together in such a way that equilibrium of forces and compatibility of displacements are satisfied at each nodal point. An example of such a process is the analysis of a continuous beam by the slope–deflection method, where the relationship between the moments and rotations within each individual span of the beam is first established, the spans then being combined together such that equilibrium of moments and compatibility of rotations are satisfied at the points of interconnection.

However, when a structure comprised of many members, such as a continuous beam containing many spans or a multistorey frame containing many bays, is being analysed, this type of approach can become very laborious and can involve the solution of a large number of simultaneous equations. Because of this, in the past much research effort has been devoted to developing analytical techniques, based on a physical appreciation of the structural behaviour, which would reduce the amount of work required to complete an analysis, and would not require the direct solution of many simultaneous equations. A prime example of such a technique is the Hardy Cross moment distribution method, in which, instead of setting up the simultaneous equations explicitly as in the slope–deflection method, the solution is accomplished in a series of convenient steps.

With the advent of the electronic digital computer, however, engineers realised that the solution of a large number of simultaneous equations no longer posed an insurmountable problem and this prompted a return to fundamental methods of analysis, such as the slope-deflection method. These methods, since they involve a number of repetitive steps, are particularly suitable for automatic computation, and they have been formulated to take maximum advantage of the capabilities of a digital computer. These so-called 'matrix methods' for analysing skeletal structures have been firmly established for a number of years.

In addition to skeletal structures, engineers are often also concerned with the analysis of continuum structures, such as deep beams, plates and slabs subjected to bending, dam walls, folded-plate and shell structures, where the structural surface is continuous instead of being composed of a number of individual components. Classical methods, such as the classical theory of plate flexure, can be applied to the analysis of these continua, but such methods have very limited fields of application because of the great difficulties that are experienced when dealing with any irregularities in structural geometry or applied loading conditions.

The finite element method represents the extension of matrix methods for skeletal structures to the analysis of continuum structures. In the finite element method, the continuum is idealised as a structure consisting of a number of individual elements connected only at nodal points, as shown in Figure 1.2. It is only in this idealisation that the method differs from the standard matrix method, as is

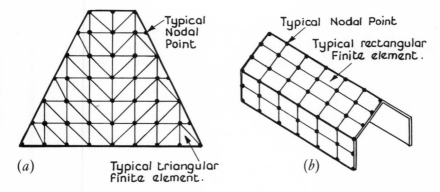

Fig. 1.2. Typical finite element idealisations of continua. *a* Dam wall. *b* Folded plate

shown later. The finite element method is extremely powerful since it enables continua with complex geometrical properties and loading conditions to be accurately analysed. The method involves extensive computations but, because of the repetitive nature of these computations, it is ideally suited for programming for a solution using a computer.

Basic structural analysis

Before commencing on a detailed discussion of the finite element method some of the relevant basic concepts of structural analysis are reviewed.

a: Fundamental requirements

Whatever the cause of the internal forces and deformations in a structure three basic conditions must be observed. These are:

    i the equilibrium of forces;
    ii the compatibility of displacements; and
    iii the laws of material behaviour.

The first condition merely requires that the internal forces balance the external applied loads. Although the use of this condition alone is sometimes sufficient to enable a statically determinate problem to be solved, the conditions of compatibility and material behaviour then being automatically satisfied, for redundant structures it yields insufficient information to enable a complete analysis to be conducted. In these circumstances the condition of compatibility must be invoked separately. Compatibility requires that the deformed structure fits together, i.e. that the deformations of the members are compatible.

Before this condition can be used it is necessary to know the relationship between load and deformation for each component of the structure. This relationship, which in problems of linear elasticity reduces to the use of Hooke's Law, is the third condition.

The use of these three conditions is a fundamental requirement of any method of structural analysis.

b: Stiffness and flexibility methods of matrix analysis

The matrix methods of structural analysis may be formulated in three different ways.

    i Stiffness (displacement) method.
    ii Flexibility (force) method.
    iii Mixed method.

The stiffness and flexibility methods differ in the order in which the two basic conditions of joint (or nodal) equilibrium and compatibility are treated. In the stiffness method, the displacement compatibility conditions are satisfied and the equations of equilibrium set up and solved to yield the unknown nodal displacements. In the flexibility method the conditions of joint equilibrium are first satisfied and the equations arising from the need for compatibility of nodal displace-ments solved to yield the unknown forces in the members. In addition to these two basic approaches, in recent years a mixed formulation involving both approaches has also been used.

c: Principle of virtual work

In later sections the *principle of virtual work* is used in deriving the stiffness properties of various elements. This principle is concerned with the relationship which exists between a set of external loads and the corresponding internal forces which together satisfy the equili-brium condition, and also with sets of joint (node) displacements and the corresponding member deformations which satisfy the condi-tions of compatibility. The principle may be stated in general terms as follows: the virtual work done by the external loads is equal to the internal virtual work absorbed by the structure. It should be emphasised that each of the systems (forces and displacements) may be either real or virtual, the only limitations being that the conditions of equilibrium and compatibility are both satisfied. The principle may be expressed in mathematical terms by the expression

$$\sum F.\delta = \int^v \sigma.\varepsilon.d \text{ (vol)} \tag{1.1}$$

where F refers to the system of external loads, δ to the deflections of the loads, σ to the system of internal forces, and ε to the internal deformations of the structure.

If the problem is one of a simple pin-jointed frame the external loads F and the deflections δ correspond to the loads at the joints and the deflections of the joints respectively, while the internal forces σ and deformations ε relate to the forces in and the extensions of the members.

For a more detailed study of the principle of virtual work reference should be made to References 1 and 2.

Objectives

The purpose of this book is to enable the reader to obtain a sound basic understanding of the finite element method. Appendix 2 contains notes on matrix algebra which must be understood before the rest of the text can be followed. In Chapter 2, the concept of a matrix analysis of a skeletal structure by the stiffness method is explained in detail and the basic ideas of the finite element method are introduced in Chapter 3, a simple one-dimensional beam element being used as an illustration. In Chapter 3 the basic steps to be followed in the derivation of the stiffness matrix for any element are established (these steps are summarised in Appendix 1). In Chapters 5, 6, 7, 8, 9 and 11 these steps are followed explicitly during the derivation of the stiffness matrix of a variety of elements which can be applied to the analysis of a large number of engineering structures. The way in which the complete analysis can best be carried out within a computer is discussed in Chapter 10, particular attention being given to techniques which economise on computer time and storage requirements. Finally, Chapters 11 and 12 are devoted to a discussion of other elements that are available, and these two chapters, together with a perusal of the publications listed in the References at the end of each chapter, and the Bibliography, should give a comprehensive picture of the power and potential of the finite element method.

Throughout·the text, the authors have aimed at simplicity of presentation and explanation with the objective of removing some of the mystique that surrounds the finite element method, so that engineers, either by developing their own computer programs or by the intelligent application of some of the excellent programs already available, will be suitably prepared to utilise the method to its full potential.

References

1. NEAL, B. G. *Structural theorems and their applications*. London, Pergamon Press, 1964.
2. WILLIAMS, A. *The analysis of indeterminate structures*. London, Macmillan & Co. Ltd, 1967.

CHAPTER TWO

The Concept of Stiffness Analysis

Introduction

The simplest structural component with which most engineers are familiar is the pin-ended tie, which has characteristics that are similar to those of the elastic spring shown in Figure 2.1. For such a component there exists a direct relationship between the force in the spring F and the displacement δ of its free end. This relationship takes the form of equation 2.1.

$$F = k\delta \tag{2.1}$$

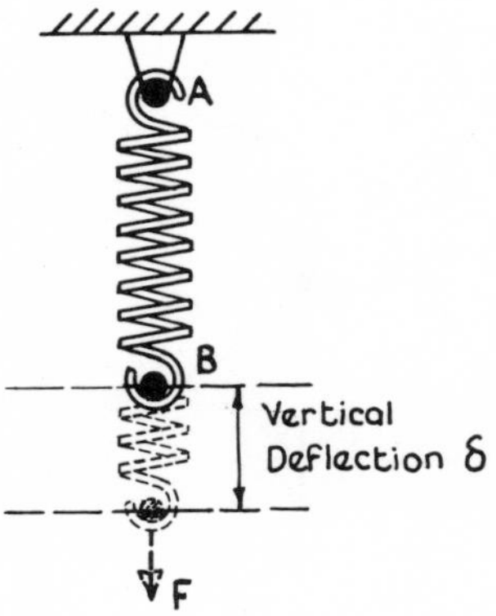

Fig. 2.1. Simple elastic spring

The quantity k is termed the stiffness of the spring and corresponds to the slope of the force–displacement diagram (Figure 2.2). Knowing the value of this stiffness and of the applied force, equation 2.1 may be inverted to give the displacement.

$$\delta = \frac{1}{k} F \qquad (2.2)$$

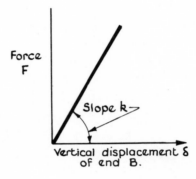

Fig. 2.2. Force–displacement relationship of a simple elastic spring

Now while the value of one displacement is sufficient to specify fully the deformed state of the simple spring shown in Figure 2.1, when dealing with more complicated structures such as the statically redundant pin-jointed structure shown in Figure 2.3, it is necessary

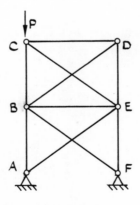

Fig. 2.3. Statically redundant pin-jointed frame

to determine the deflections of joints B, C, D and E in order to be able to evaluate the stresses occurring in the members. In the following discussion, points such as B, C, D and E are termed 'nodes'. Assume that for a complete structure it is also possible to derive a quantity similar to k in equation 2.1. For this case it is necessary to rewrite equation 2.1 in matrix form as shown in equation 2.3 since, as is shown later in this chapter, a number of simple members are now inter-connected at a number of nodes and the resulting force–displacement action of the structure can only be described by means of a series of simultaneous equations. (A brief outline of the matrix algebra required is given in Appendix 2.) In equation 2.3, the quantities $\{F\}$ and $\{\delta\}$ are vectors of nodal loads and nodal displacements respectively. Then

$$\{F\} = [K]\{\delta\} \qquad (2.3)$$

where the quantity $[K]$ is the 'stiffness' of the complete structure.

This concept is the basis of the stiffness method of analysis. The quantity $[K]$ is termed the stiffness matrix for the structure and relates the applied nodal forces $\{F\}$ to the unknown nodal displacements $\{\delta\}$. For the very simple structure shown in Figure 2.1, which has only one possible displacement, this matrix is of order 1×1 and the vectors $\{F\}$ and $\{\delta\}$ contain only one term each.

Later in this text details of how this matrix may be assembled for a complete structure are given. First the form which the stiffness matrix takes for the case of simple pin-jointed members is examined, since a full understanding of the matrix analysis of structures consisting of simple line members provides the key to the whole finite element method of analysis.

Stiffness matrix for single elastic spring

The spring shown in Figure 2.1 can have only one possible displacement δ. However, were this spring part of a structure each of its ends would be connected to other parts of the structure and could be displaced. As stated earlier these points of attachment to the other parts of the structure are called nodes. A simple spring element, therefore, possesses two nodes, each of which may deflect and to each of which a force may be applied.

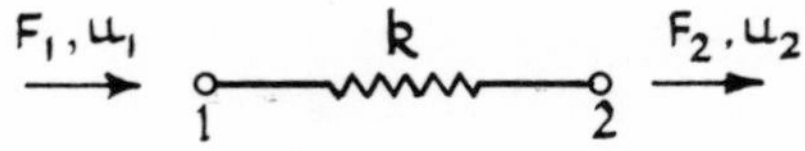

Fig. 2.4. Equivalent spring element for a pin-jointed tie

In Figure 2.4, F_1 and u_1 are the force and displacement acting axially at end 1, and F_2 and u_2 the force and displacement acting axially at end 2. The force vector for the spring is therefore

$$\left\{ \begin{array}{c} F_1 \\ F_2 \end{array} \right\}$$

and the displacement vector is

$$\left\{ \begin{array}{c} u_1 \\ u_2 \end{array} \right\}$$

The stiffness matrix for the spring is therefore of order 2×2 and equation 2.3 takes the form

$$\left\{ \begin{array}{c} F_1 \\ F_2 \end{array} \right\} = \left[\begin{array}{cc} k_{11} & k_{12} \\ k_{21} & k_{22} \end{array} \right] \left\{ \begin{array}{c} u_1 \\ u_2 \end{array} \right\} \tag{2.4}$$

where the individual terms in the stiffness matrix are as yet undetermined. These terms are obtained by permitting the element to adopt each independent mode of deformation in turn and determining the relationship between this deformation (in this case an extension) and the nodal forces.

In the following discussion the sign convention illustrated in Figure 2.5 is adopted for both forces and displacements when

Fig. 2.5. Sign convention

dealing with linear springs in series. First it is assumed that only end 1 can deflect, node 2 being fixed, as shown in Figure 2.6a. This case is identical to that of a single spring already described. Consequently, the force and displacement at node 1 are related by the equation $F_{1a} = ku_1$ where u_1 is the displacement at end A. Equilibrium of the forces acting on the spring then requires that

$$F_{1a} + F_{2a} = 0$$
$$F_{2a} = -F_{1a} = -ku_1$$

It should be noted that continuity of displacements is automatically satisfied for this simple spring.

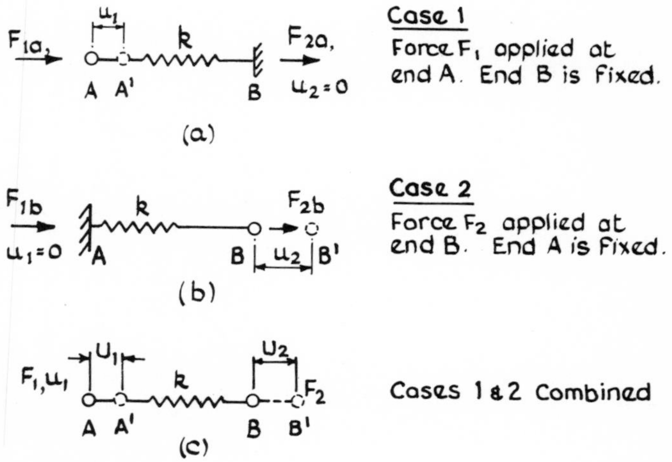

Case 1
Force F_1 applied at end A. End B is fixed.

Case 2
Force F_2 applied at end B. End A is fixed.

Cases 1 & 2 Combined

Fig. 2.6. Possible deflected states for the spring AB.

If the situation is now reversed by fixing node 1 in its initial position and allowing node 2 to deflect under the action of a force F_2 applied at end B (Figure 2.6b),

$$F_{2b} = ku_2 = -F_{1b}$$

To obtain the relationship between the forces F_1 and F_2 and the displacements u_1 and u_2 for the case when both joints are forced to displace, as shown in Figure 2.6c, using the principle of superposition by combining the load systems shown in Figures 2.6a and 2.6b,

Total force acting at node 1: $F_1 = F_{1a} + F_{1b}$
Total force acting at node 2: $F_2 = F_{2a} + F_{2b}$
or
$$F_1 = ku_1 - ku_2$$
$$F_2 = -ku_1 + ku_2$$

It may be noted that these equations can easily be written in the matrix form of equation 2.3 as follows.

$$\begin{Bmatrix} F_1 \\ F_2 \end{Bmatrix} = \begin{bmatrix} k & -k \\ -k & k \end{bmatrix} \begin{Bmatrix} u_1 \\ u_2 \end{Bmatrix} \qquad (2.5a)$$

Thus the stiffness matrix $[K^e]$ for the single spring element is given by equation 2.5, the suffix e being used to indicate that the matrix is for a single element.

$$[K^e] = \begin{bmatrix} k & -k \\ -k & k \end{bmatrix} \qquad (2.5b)$$

An important property of the stiffness matrix of a single element and indeed of a complete structure may be observed in equation 2.5. The stiffness matrix is symmetrical, the coefficient k_{21} being equal to the coefficient k_{12}, as would be expected as a consequence of the reciprocal theorem.

It may also be noted that the matrix for the element given in equation 2.5 is singular, i.e. the value of its determinant is zero.

Stiffness matrix for assembly of springs

Having seen how to form the stiffness matrix for a single structural element, albeit a very simple one, the next step is to decide how these matrices can be combined together to form the stiffness matrix of a structure composed of several such elements. In the first instance a process similar to that described above is used to obtain $[K]$ for the assembly of two co-linear springs shown in Figure 2.7 which, on

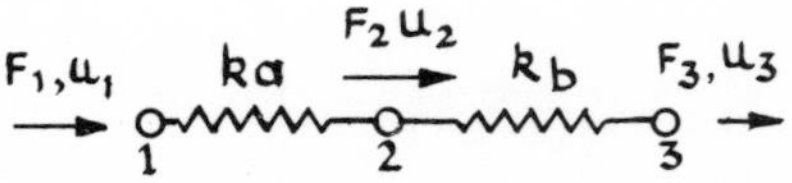

Fig. 2.7. Assembly of spring elements

inspection, indicates how $[K]$ for the structure may be formed from the stiffness matrices $[K^e]$ of the individual elements.

Case 1
Proceeding as in the previous example, u_2 and u_3 are first set equal to

zero, allowing only node 1 to deflect (Figure 2.8a). For this case, the relationship between F_1 and u_1 becomes

$$F_1 = k_a u_1$$

Considering spring AB, then from the laws of statics

$$F_2 = -F_1$$

No force can exist at node 3, since both u_2 and u_3 are specified as zero. Therefore

$$F_3 = 0$$

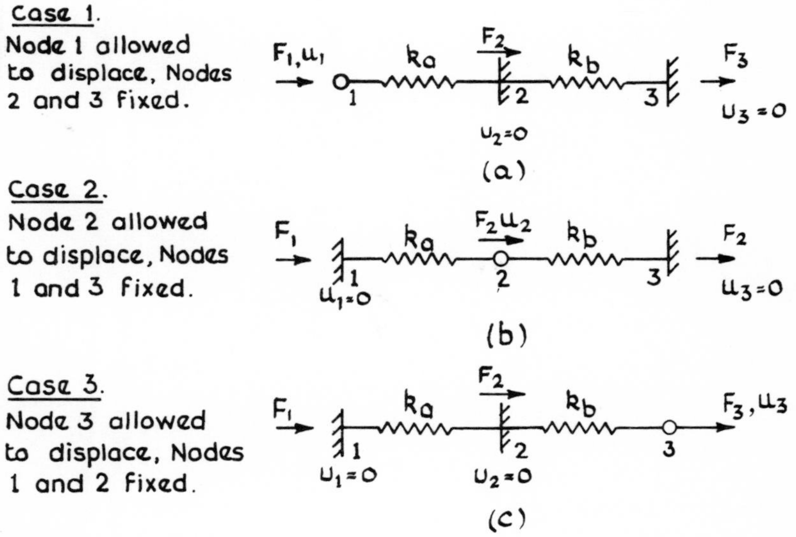

Case 1.
Node 1 allowed to displace, Nodes 2 and 3 fixed.

Case 2.
Node 2 allowed to displace, Nodes 1 and 3 fixed.

Case 3.
Node 3 allowed to displace, Nodes 1 and 2 fixed.

Fig. 2.8. Possible deflected states for assembly of springs

Case 2

Both u_1 and u_3 are now set equal to zero (Figure 2.8b), it being noted that in this case continuity of displacements at node 2 requires that each spring deflects by the same amount. Thus the force at node 2 consists of two components, $k_a u_2$ (the force needed to extend spring AB) and $k_b u_2$ (the force needed to compress spring BC). Therefore

$$F_2 = (k_a + k_b)u_2$$

Considering the equilibrium of spring AB,

$$F_1 = -k_a u_2$$

and considering the equilibrium of spring BC

$$F_3 = -k_b u_2$$

Case 3
Finally, u_1 and u_2 are set equal to zero (Figure 2.8c), and by analogy with the first case,

$$F_3 = k_b u_3$$
$$F_2 = -F_3$$

Clearly $F_1 = 0$ because nodes A and B do not move.

Combined action
It remains to recast the results obtained in the form of equation 2.3, noting that in this case equation 2.3 takes the form

$$\begin{Bmatrix} F_1 \\ F_2 \\ F_3 \end{Bmatrix} = \begin{bmatrix} k_{11} & k_{12} & k_{13} \\ k_{21} & k_{22} & k_{23} \\ k_{31} & k_{32} & k_{33} \end{bmatrix} \begin{Bmatrix} u_1 \\ u_2 \\ u_3 \end{Bmatrix} \tag{2.6}$$

Since linear elastic behaviour is being considered the principle of superposition may be used and the three loading cases added. Thus

		Case 1	*Case 2*	*Case 3*
(Total force acting at A)	$F_1 =$	$+k_a u_1$	$-k_a u_2$	0
(Total force acting at B)	$F_2 =$	$-k_a u_1$	$+k_a u_2 + k_b u_2$	$-k_b u_3$
(Total force acting at C)	$F_3 =$	0	$-k_b u_2$	$+k_b u_3$

Writing these equations in matrix form gives

$$[K] = \begin{bmatrix} k_a & -k_a & 0 \\ -k_a & k_a + k_b & -k_b \\ 0 & -k_b & k_b \end{bmatrix} \tag{2.7}$$

Note that once again the stiffness matrix $[K]$ is symmetric.

Although the assembly of the matrix $[K]$ of equation 2.7 is not difficult in this particular case, it would prove extremely tedious if the structure comprised a large number of springs. Since the form of $[K^e]$ for a single spring is already known (equation 2.5), it may be asked whether it is possible to obtain the stiffness matrix for the structure $[K]$ from the stiffness matrices of the individual elements $[K^e]$, thus removing the necessity of considering every possible state of displacement. Although equations 2.5 and 2.7 possess certain similarities, the method of doing this may not be immediately obvious.

First $[K^e]$ for each of the constituent elements is written down.

Element 1

$$\begin{Bmatrix} F_1 \\ F_2 \end{Bmatrix} = \begin{bmatrix} k_a & -k_a \\ -k_a & k_a \end{bmatrix} \begin{Bmatrix} u_1 \\ u_2 \end{Bmatrix}$$

Element 2

$$\begin{Bmatrix} F_2 \\ F_3 \end{Bmatrix} = \begin{bmatrix} k_b & -k_b \\ -k_b & k_b \end{bmatrix} \begin{Bmatrix} u_2 \\ u_3 \end{Bmatrix}$$

Although the two $[K^e]$ matrices are of the same order, they may not be added directly since they relate to different sets of displacements. However, by inserting rows and columns of zeros, both may be expanded in such a way that they each relate to the three displacements u_1, u_2 and u_3 thus,

$$\begin{Bmatrix} F_1 \\ F_2 \\ F_3 \end{Bmatrix} = \begin{bmatrix} k_a & -k_a & 0 \\ -k_a & k_a & 0 \\ 0 & 0 & 0 \end{bmatrix} \begin{Bmatrix} u_1 \\ u_2 \\ u_3 \end{Bmatrix}$$

$$\begin{Bmatrix} F_1 \\ F_2 \\ F_3 \end{Bmatrix} = \begin{bmatrix} 0 & 0 & 0 \\ 0 & k_b & -k_b \\ 0 & -k_b & k_b \end{bmatrix} \begin{Bmatrix} u_1 \\ u_2 \\ u_3 \end{Bmatrix}$$

The rule for matrix addition (see Appendix 2), may now be used to obtain

$$\begin{Bmatrix} F_1 \\ F_2 \\ F_3 \end{Bmatrix} = \begin{bmatrix} k_a & -k_a & 0 \\ -k_a & k_a+k_b & -k_b \\ 0 & -k_b & k_b \end{bmatrix} \begin{Bmatrix} u_1 \\ u_2 \\ u_3 \end{Bmatrix} \tag{2.8}$$

This sequence of operations is identical to the use of superposition to obtain equations 2.6 except that the displacements of the structure are considered element by element instead of node by node. Reference to equation 2.8 shows that the stiffnesses of the individual elements have been added together to obtain the stiffness of the structure, due attention being paid to the forces and displacements to which the actual stiffness terms relate. For a member connecting nodes i and j it has been shown that there are direct stiffness terms k_{ii} and k_{jj} on the leading diagonal and terms k_{ij} and k_{ji} $(=k_{ij})$ in the off-diagonal position. Defining the diagonal and off-diagonal terms in $[K^e]$ as direct and indirect stiffnesses respectively, the following rules for the direct formation of $[K]$ may be formulated.

i The term in location ii consists of the sum of the direct stiffnesses of all the elements meeting at node i.

ii The term in location ij consists of the sum of the indirect stiffnesses relating to nodes i and j of all the elements joining node i to node j.

In the above simple example there is only a single term in each of the off-diagonal positions, since each pair of nodes is joined by only one element. However, when more complex structures involving elements possessing more than two nodes are dealt with, the second rule becomes equally as important as the first.

Solution procedure

The matrix of equation 2.7 is, in fact, singular. Mathematically this is equivalent to saying that its determinant vanishes, i.e. is equal to zero, and therefore, its inverse does not exist. This means that the associated set of simultaneous equations for the unknown displacements cannot be solved! However, a perfectly simple physical explanation exists for this occurrence—the structure has not been secured to the ground! As the system stands, no limitation has been placed on any of the displacements u_1, u_2 and u_3. Therefore, the application of any form of external loading will result in the system moving as a rigid body. This situation can be remedied and the problem rendered solvable simply by specifying sufficient boundary conditions to prevent the structure moving as a rigid body. There-fore assume node 1 to be fixed ($u_1 = 0$); then equation 2.8 can be rewritten in partitioned form as shown in equation 2.8a.

$$\begin{Bmatrix} F_1 \\ \hline F_2 \\ F_3 \end{Bmatrix} = \begin{bmatrix} k_a & -k_a & 0 \\ \hline -k_a & k_a+k_b & -k_b \\ 0 & -k_b & k_b \end{bmatrix} \begin{Bmatrix} u_1 = 0 \\ \hline u_2 \\ u_3 \end{Bmatrix} \qquad (2.8a)$$

This equation therefore contains an unknown reaction F_1 and two unknown displacements u_2 and u_3, F_2 and F_3 being known applied loads. Using the standard matrix rules,

$$\{F_1\} = k_a\{u_1 = 0\} + [-k_a \quad 0]\begin{Bmatrix} u_2 \\ u_3 \end{Bmatrix}$$

$$\begin{Bmatrix} F_2 \\ F_3 \end{Bmatrix} = \begin{bmatrix} -k_a \\ 0 \end{bmatrix}\{u_1 = 0\} + \begin{bmatrix} k_a+k_b & -k_b \\ -k_b & k_b \end{bmatrix}\begin{Bmatrix} u_2 \\ u_3 \end{Bmatrix}$$

Noting that u_1 is zero,

$$\{F_1\} = [-k_a \quad 0]\begin{Bmatrix} u_2 \\ u_3 \end{Bmatrix} \tag{2.9a}$$

$$\begin{Bmatrix} F_2 \\ F_3 \end{Bmatrix} = \begin{bmatrix} k_a+k_b & -k_b \\ -k_b & k_b \end{bmatrix}\begin{Bmatrix} u_2 \\ u_3 \end{Bmatrix} \tag{2.9b}$$

Equation 2.9b consists of two equations in the two unknowns u_2 and u_3 and may be solved to yield u_2 and u_3 which, when substituted in equation 2.9a, then give the value of the unknown reaction F_1.

It should be noted that equation 2.9b may be obtained directly from equation 2.8a simply by deleting the rows and columns of $[K]$ corresponding to the zero displacements.

Once the displacements have been obtained, the internal forces in the elements may be determined with the aid of the force–displacement relations of the springs. If p_a is the internal force in spring AB, then $p_a = k_a \times$ (change in length of spring AB) or

$$p_a = k_a(u_2 - u_1)$$

Similarly

$$p_b = k_b(u_3 - u_2)$$

This completes the solution process.

Stage	Operation
1	Form each element stiffness matrix $[K^e]$.
2	Assemble the structural stiffness matrix $[K]$ from the individual element stiffness matrices $[K^e]$.
3	Apply the boundary conditions.
4	Solve for the displacements and then, if required, the reactions.
5	Use the force–displacement relations of the element to obtain the element forces.

Table 2.1. Sequence of operations involved in stiffness analysis

It is useful to list the steps in the analysis since these remain essentially similar whatever the problem and the types of element being used (Table 2.1, p. 17).

Simple example

A simple numerical example illustrates the above procedure. The three springs shown in Figure 2.9 are collinear and nodes 1 and 4 are fixed. If axial loads of 10 kN and 20 kN are applied at nodes 2 and 3 respectively as shown, determine the displacements at nodes 2 and 3.

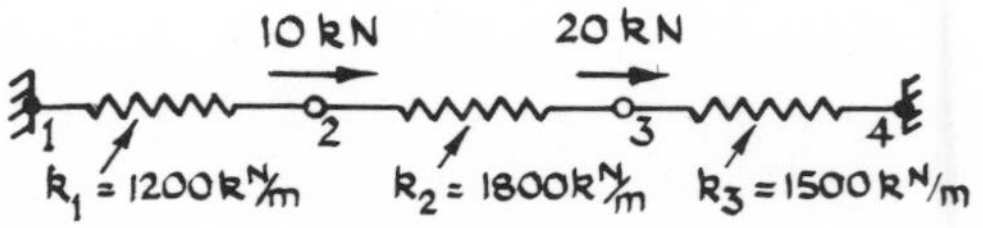

Fig. 2.9.

The boundary conditions are $u_1 = u_4 = 0$, u_2 and u_3 are the unknown deflections; F_1 and F_4 are the unknown reactions; $F_2 = 10$ kN and $F_3 = 20$ kN.

Stiffness matrix for spring 1–2

$$\begin{Bmatrix} F_1 \\ F_2 \end{Bmatrix} = \begin{bmatrix} 1200 & -1200 \\ -1200 & 1200 \end{bmatrix} \begin{Bmatrix} u_1 \\ u_2 \end{Bmatrix}$$

Stiffness matrix for spring 2–3

$$\begin{Bmatrix} F_2 \\ F_3 \end{Bmatrix} = \begin{bmatrix} 1800 & -1800 \\ -1800 & 1800 \end{bmatrix} \begin{Bmatrix} u_2 \\ u_3 \end{Bmatrix}$$

Stiffness matrix for spring 3–4

$$\begin{Bmatrix} F_3 \\ F_4 \end{Bmatrix} = \begin{bmatrix} 1500 & -1500 \\ -1500 & 1500 \end{bmatrix} \begin{Bmatrix} u_3 \\ u_4 \end{Bmatrix}$$

Assembling these matrices to form the overall matrix for the structure

$$
\begin{Bmatrix} F_1 = ? \\ F_2 = 10 \\ F_3 = 20 \\ F_4 = ? \end{Bmatrix} =
\begin{bmatrix}
1200 & -1200 & 0 & 0 \\
-1200 & \begin{matrix} 1200 \\ +1800 \end{matrix} & -1800 & 0 \\
0 & -1800 & \begin{matrix} 1800 \\ +1500 \end{matrix} & -1500 \\
0 & 0 & -1500 & 1500
\end{bmatrix}
\begin{Bmatrix} u_1 = 0 \\ u_2 = ? \\ u_3 = ? \\ u_4 = 0 \end{Bmatrix}
\tag{2.10}
$$

Summing the terms and crossing out the rows and columns associated with the zero displacements u_1 and u_4,

$$
\begin{Bmatrix} F_1 \\ 10 \\ 20 \\ F_4 \end{Bmatrix} =
\begin{bmatrix}
\cancel{1200} & \cancel{-1200} & \cancel{0} & \cancel{0} \\
-\cancel{1200} & 3000 & -1800 & \cancel{0} \\
\cancel{0} & -1800 & 3300 & \cancel{-1500} \\
\cancel{0} & \cancel{0} & \cancel{-1500} & \cancel{1500}
\end{bmatrix}
\begin{Bmatrix} 0 \\ u_2 \\ u_3 \\ 0 \end{Bmatrix}
$$

To determine the unknown displacements u_2 and u_3 the following sub-matrix must be solved.

$$
\begin{Bmatrix} 10 \\ 20 \end{Bmatrix} =
\begin{bmatrix} 3000 & -1800 \\ -1800 & 3300 \end{bmatrix}
\begin{Bmatrix} u_2 \\ u_3 \end{Bmatrix}
$$

These two simultaneous equations give $u_2 = 0 \cdot 0103603$ m and $u_3 = 0 \cdot 0117117$ m. The values of the four displacements u_1, u_2, u_3 and u_4 are now known and the magnitudes of the unknown reactions can be determined by multiplying out the original matrix (equation 2.10).

$$
F_1 = 1200u_1 - 1200u_2 + 0u_3 + 0u_4
$$

hence

$$
F_1 = -1200u_2 = -12 \cdot 432 \text{ kN}
$$
$$
F_4 = 0u_1 + 0u_2 - 1500u_3 + 1500u_4
$$
$$
= -1500 \times 0 \cdot 0117117 = -17 \cdot 567 \text{ kN.}
$$

(Check $F_1 + F_4 = -29 \cdot 999 \simeq -30$ kN)

Application to frameworks

The method of analysing a series of collinear springs or rods which are assumed to be only capable of carrying axial forces has now been shown. For a uniform pin-jointed bar, the value of its stiffness k can be obtained from the standard stress–strain relationship

$$F_1 = \frac{AE}{L} u_1$$

where F_1 is the tension in the member and u_1 is the extension of the member. Therefore $k = AE/L$. Thus for a uniform section bar equation 2.4 takes the form

$$\begin{Bmatrix} F_1 \\ F_2 \end{Bmatrix} = \frac{AE}{L} \begin{bmatrix} 1 & -1 \\ -1 & 1 \end{bmatrix} \begin{Bmatrix} u_1 \\ u_2 \end{Bmatrix} \tag{2.11}$$

Equation 2.11 represents the relationship between the forces acting on the ends of the member and the displacements occurring at the nodes, written in terms of the member co-ordinates. However, since structural frameworks usually consist of members set at various angles to one another (e.g. Figure 2.3) it is necessary to make due allowance for this fact. When developing the overall stiffness matrix for a complete structure it is necessary to write the stiffness matrix for each element of the structure, not in terms of its own member co-ordinates, but in the global co-ordinate system adopted for the complete structure.

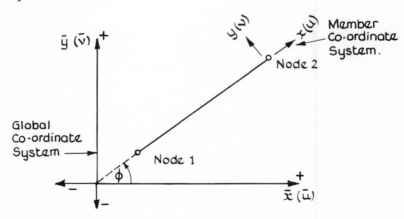

Fig. 2.10. Co-ordinate systems

In the following discussion the sign convention shown in Figure 2.10 is employed. It should be noted that the angle ϕ is positive when measured anti-clockwise from the global $\bar{x}$-axis.

In Figure 2.10, a pin-jointed member 1–2 is inclined at an angle ϕ to the global system. Axes x and y relate to the local member system and $\bar{x}$ and $\bar{y}$ to the global system, the respective displacements being u and v and $\bar{u}$ and $\bar{v}$ and the forces F_x, F_y, $\bar{F}_x$ and $\bar{F}_y$. Since an axial displacement u of a member generally possesses both a $\bar{u}$ and a $\bar{v}$ component in the global system it is necessary to expand equation 2.11 as shown below.

$$\begin{Bmatrix} F_{x1} \\ F_{y1} \\ F_{x2} \\ F_{y2} \end{Bmatrix} = \frac{AE}{L} \begin{bmatrix} 1 & 0 & -1 & 0 \\ 0 & 0 & 0 & 0 \\ -1 & 0 & 1 & 0 \\ 0 & 0 & 0 & 0 \end{bmatrix} \begin{Bmatrix} u_1 \\ v_1 \\ u_2 \\ v_2 \end{Bmatrix} \qquad (2.11a)$$

Since a pin-jointed member can only carry an axial load, $F_{y1} = F_{y2} = 0$. It may be noted from Figure 2.11 that the local and global system of forces at node 1 may be related by the expressions

$$F_{x1} = \bar{F}_{x1} \cos \phi + \bar{F}_{y1} \sin \phi$$
$$F_{y1} = -\bar{F}_{x1} \sin \phi + \bar{F}_{y1} \cos \phi$$

and similar expressions apply for node 2.

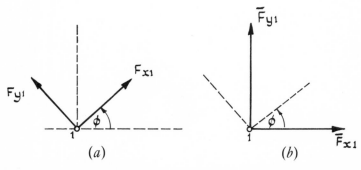

Fig. 2.11. *a* Local member forces acting at node 1 when expressed in terms of member co-ordinates. *b* Local member forces at node 1 when expressed in global co-ordinates

Writing μ for $\sin \phi$ and λ for $\cos \phi$, for the complete system of forces

$$\begin{Bmatrix} F_{x1} \\ F_{y1} \\ F_{x2} \\ F_{y2} \end{Bmatrix} = \begin{bmatrix} \lambda & \mu & 0 & 0 \\ -\mu & \lambda & 0 & 0 \\ 0 & 0 & \lambda & \mu \\ 0 & 0 & -\mu & \lambda \end{bmatrix} \begin{Bmatrix} \overline{F}_{x1} \\ \overline{F}_{y1} \\ \overline{F}_{x2} \\ \overline{F}_{y2} \end{Bmatrix} \qquad (2.12)$$

or

$$\{F\} = [T]\{\overline{F}\}$$

where $[T]$ is called the transformation matrix. A very useful property of the transformation matrix is that its inverse is equal to its transpose, i.e.

$$[T]^{-1} = [T]^{T}$$

This is demonstrated in the basic matrix notes given in Appendix 2.

The relationship which exists between the two sets of displacements is similar to that existing between the two sets of forces, namely

$$\{\delta\} = [T]\{\overline{\delta}\} \qquad (2.13)$$

All the necessary data to establish the stiffness matrix $[\overline{K}^e]$ for the member in global co-ordinates is now available. The basic force–displacement relationship for the element given by equation 2.11a states that $\{F\}=[K^e]\{\delta\}$ and substituting for $\{F\}$ from equation 2.12,

$$[T]\{\overline{F}\} = [K^e]\{\delta\}$$

Pre-multiplying both sides by $[T]^{-1}$ and using the fact that $[T]^{-1} = [T]^{T}$,

$$[T]^{T}[T]\{\overline{F}\} = [T]^{T}[K^e]\{\delta\}$$

Therefore

$$\{\overline{F}\} = [T]^{T}[K^e]\{\delta\} \qquad (2.14)$$

From equation 2.13, $\{\delta\}=[T]\{\overline{\delta}\}$ and hence on substituting for $\{\delta\}$ into equation 2.14,

$$\{\overline{F}\} = [T]^{T}[K^e][T]\{\overline{\delta}\} = [\overline{K}^e]\{\overline{\delta}\}$$

From this it can be seen that the stiffness matrix $[\overline{K}^e]$ for the member written in global co-ordinates is obtained from equation 2.15, in which $[K^e]$ is the element stiffness matrix in local co-ordinates.

$$[\overline{K}^e] = [T]^T[K^e][T] \tag{2.15}$$

In Appendix 2 the multiplication of these three matrices is carried out and the explicit form for $[\overline{K}^e]$ is shown to become

$$[\overline{K}^e] = \frac{AE}{L}\begin{bmatrix} \lambda^2 & \lambda\mu & -\lambda^2 & -\lambda\mu \\ \lambda\mu & \mu^2 & -\lambda\mu & -\mu^2 \\ -\lambda^2 & -\lambda\mu & \lambda^2 & \lambda\mu \\ -\lambda\mu & -\mu^2 & \lambda\mu & \mu^2 \end{bmatrix} \tag{2.16}$$

It should be noted that $[\overline{K}^e]$ is a symmetric matrix consisting of four sub-matrices.

$$[\overline{K}^e] = \frac{AE}{L}\left[\begin{array}{cc|cc} \lambda^2 & \lambda\mu & -\lambda^2 & -\lambda\mu \\ \lambda\mu & \mu^2 & -\lambda\mu & -\mu^2 \\ \hline -\lambda^2 & -\lambda\mu & \lambda^2 & \lambda\mu \\ -\lambda\mu & -\mu^2 & \lambda\mu & \mu^2 \end{array}\right] = \frac{AE}{L}\begin{bmatrix} k^e & -k^e \\ -k^e & k^e \end{bmatrix}$$

where

$$[k^e] = \begin{bmatrix} \lambda^2 & \lambda\mu \\ \lambda\mu & \mu^2 \end{bmatrix}$$

It may be observed that the stiffness matrix $[\overline{K}^e]$ written in global co-ordinates is similar in form to equation 2.5b. It is important to recognize that if the stiffness matrix is first assembled in local co-ordinates it must then be transformed into the global co-ordinate system before commencing the assembly process.

Having obtained the displacements of the nodes in the global system it is convenient to revert to the local system to evaluate the element forces. Performing the algebra leads to

$$\sigma_{ij} = \left(\frac{AE}{L}\right)_{ij}[\lambda \quad \mu]_{ij}\begin{Bmatrix} (\bar{u}_j - \bar{u}_i) \\ (\bar{v}_j - \bar{v}_i) \end{Bmatrix} \tag{2.17}$$

where equation 2.17 relates to a general member connecting nodes i and j in the global system.

The complete process is now demonstrated with the aid of the simple example illustrated in Figure 2.12. The three-member steel truss is supported on non-yielding supports at 1 and 3 and all members have the same cross-sectional area A.

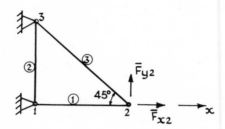

Fig. 2.12. Simple example of pin-jointed framework

Since there are three nodes with two degrees of freedom at each node, the vectors of applied loads and displacements each contain six terms as shown below.

$$\{\overline{F}\} = \begin{Bmatrix} \overline{F}_{x1} \\ \overline{F}_{y1} \\ \overline{F}_{x2} \\ \overline{F}_{y2} \\ \overline{F}_{x3} \\ \overline{F}_{y3} \end{Bmatrix} \qquad \{\overline{\delta}\} = \begin{Bmatrix} \overline{u}_1 = 0 \\ \overline{v}_1 = 0 \\ \overline{u}_2 \\ \overline{v}_2 \\ \overline{u}_3 = 0 \\ \overline{v}_3 = 0 \end{Bmatrix}$$

The expression for $[\overline{K}^e]$ for a typical element has already been given in equation 2.16. Before assembling $[\overline{K}^e]$ for each element, however, the direction cosines λ and μ for each element must be determined. Remembering that ϕ is measured anticlockwise from the positive x-axis to member ij these become

$$
\begin{array}{lllll}
& & & \cos\phi & \sin\phi \\
\text{Member 1:} & \phi = & 0° & \lambda = 1 & \mu = 0 \\
\text{Member 2:} & \phi = & 90° & \lambda = 0 & \mu = 1 \\
\text{Member 3:} & \phi = & 135° & \lambda = -\dfrac{1}{\sqrt{2}} & \mu = \dfrac{1}{\sqrt{2}}
\end{array}
$$

Since the terms in $[K^e]$ involve λ^2, μ^2 and $\mu\lambda$, a change in angle from ϕ to $(\phi+\pi)$, which causes both μ and λ to change sign, does not affect the signs of the terms in the stiffness matrix. For example, with respect to member 3, $\phi=135°$ if measured anti-clockwise at node 2, or $315°$ if measured anti-clockwise at node 3. However, by substituting into equation 2.16 it is clear that $[K^e]$ remains unchanged.

For the three members the $[\overline{K}^e]$ matrices therefore become

Member 1:

$$
[\overline{K}_1{}^e] = \frac{AE}{L}
\begin{bmatrix}
1 & 0 & -1 & 0 \\
0 & 0 & 0 & 0 \\
-1 & 0 & 1 & 0 \\
0 & 0 & 0 & 0
\end{bmatrix}
$$

Member 2:

$$
[\overline{K}_2{}^e] = \frac{AE}{L}
\begin{bmatrix}
0 & 0 & 0 & 0 \\
0 & 1 & 0 & -1 \\
0 & 0 & 0 & 0 \\
0 & -1 & 0 & 1
\end{bmatrix}
$$

Member 3:

$$
[\overline{K}_3{}^e] = \frac{AE}{\sqrt{2}\,L}
\begin{bmatrix}
\dfrac{1}{2} & \dfrac{-1}{2} & \dfrac{-1}{2} & \dfrac{1}{2} \\[2mm]
\dfrac{-1}{2} & \dfrac{1}{2} & \dfrac{1}{2} & \dfrac{-1}{2} \\[2mm]
\dfrac{-1}{2} & \dfrac{1}{2} & \dfrac{1}{2} & \dfrac{-1}{2} \\[2mm]
\dfrac{1}{2} & \dfrac{-1}{2} & \dfrac{-1}{2} & \dfrac{1}{2}
\end{bmatrix}
$$

These matrices are added together to give the stiffness matrix for the structure, as shown symbolically in the following diagram.

$$\begin{Bmatrix} \bar{F}_{x1} \\ \bar{F}_{y1} \\ \bar{F}_{x2} \\ \bar{F}_{y2} \\ \bar{F}_{x3} \\ \bar{F}_{y3} \end{Bmatrix} = \begin{bmatrix} ①+② & ①+② & ① & ① & ② & ② \\ ①+② & ①+② & ① & ① & ② & ② \\ ① & ① & ①+③ & ①+③ & ③ & ③ \\ ① & ① & ①+③ & ①+③ & ③ & ③ \\ ② & ② & ③ & ③ & ②+③ & ②+③ \\ ② & ② & ③ & ③ & ②+③ & ②+③ \end{bmatrix} \begin{Bmatrix} \bar{u}_1 \\ \bar{v}_1 \\ \bar{u}_2 \\ \bar{v}_2 \\ \bar{u}_3 \\ \bar{v}_3 \end{Bmatrix}$$

where ① represents the terms for Member 1 (having nodes 1 and 2), ② represents the terms for Member 2 (having nodes 1 and 3), and ③ represents the terms for Member 3 (having nodes 3 and 2). Using the actual values for the individual matrices, the overall stiffness matrix $[\bar{K}]$ is

$$[\bar{K}] = \frac{AE}{L} \begin{bmatrix} 1 & 0 & -1 & 0 & 0 & 0 \\ 0 & 1 & 0 & 0 & 0 & -1 \\ -1 & 0 & 1+\dfrac{1}{2\sqrt{2}} & \dfrac{-1}{2\sqrt{2}} & \dfrac{-1}{2\sqrt{2}} & \dfrac{1}{2\sqrt{2}} \\ 0 & 0 & \dfrac{-1}{2\sqrt{2}} & \dfrac{1}{2\sqrt{2}} & \dfrac{1}{2\sqrt{2}} & \dfrac{-1}{2\sqrt{2}} \\ 0 & 0 & \dfrac{-1}{2\sqrt{2}} & \dfrac{1}{2\sqrt{2}} & \dfrac{1}{2\sqrt{2}} & \dfrac{-1}{2\sqrt{2}} \\ 0 & -1 & \dfrac{1}{2\sqrt{2}} & \dfrac{-1}{2\sqrt{2}} & \dfrac{-1}{2\sqrt{2}} & 1+\dfrac{1}{2\sqrt{2}} \end{bmatrix}$$

At this stage a check should be made to see that $[\bar{K}]$ is symmetrical and that all the diagonal terms are positive.

Noting that nodes 1 and 3 are fixed, the governing matrix equation may, for convenience, be rewritten in the following form.

$$
\begin{Bmatrix} \bar{F}_{x2} \\ \bar{F}_{y2} \\ \bar{F}_{x1} \\ \bar{F}_{y1} \\ \bar{F}_{x3} \\ \bar{F}_{y3} \end{Bmatrix} = \frac{AE}{L}
\begin{bmatrix}
1+\dfrac{1}{2\sqrt{2}} & \dfrac{-1}{2\sqrt{2}} & -1 & 1 & \dfrac{-1}{2\sqrt{2}} & \dfrac{1}{2\sqrt{2}} \\[2mm]
\dfrac{-1}{2\sqrt{2}} & \dfrac{1}{2\sqrt{2}} & 0 & 0 & \dfrac{1}{2\sqrt{2}} & \dfrac{-1}{2\sqrt{2}} \\[2mm]
-1 & 0 & 1 & 0 & 0 & 0 \\[2mm]
0 & 0 & 0 & 1 & 0 & -1 \\[2mm]
\dfrac{-1}{2\sqrt{2}} & \dfrac{1}{2\sqrt{2}} & 0 & 0 & \dfrac{1}{2\sqrt{2}} & \dfrac{-1}{2\sqrt{2}} \\[2mm]
\dfrac{1}{2\sqrt{2}} & \dfrac{-1}{2\sqrt{2}} & 0 & -1 & \dfrac{-1}{2\sqrt{2}} & 1+\dfrac{1}{2\sqrt{2}}
\end{bmatrix}
\begin{Bmatrix} \bar{u}_2 \\ \bar{v}_2 \\ \bar{u}_1 = 0 \\ \bar{v}_1 = 0 \\ \bar{u}_3 = 0 \\ \bar{v}_3 = 0 \end{Bmatrix}
$$

To determine the unknown displacements $\bar{u}_2$ and $\bar{v}_2$ only the part of this equation relating directly to these two displacements is considered.

$$
\begin{Bmatrix} \bar{F}_{x2} \\ \bar{F}_{y2} \end{Bmatrix} = \frac{AE}{L}
\begin{bmatrix}
1+\dfrac{1}{2\sqrt{2}} & \dfrac{-1}{2\sqrt{2}} \\[2mm]
\dfrac{-1}{2\sqrt{2}} & \dfrac{1}{2\sqrt{2}}
\end{bmatrix}
\begin{Bmatrix} \bar{u}_2 \\ \bar{v}_2 \end{Bmatrix}
$$

This may be inverted to yield

$$
\begin{Bmatrix} \bar{u}_2 \\ \bar{v}_2 \end{Bmatrix} = \frac{L}{AE}
\begin{bmatrix}
1 & 1 \\
1 & 1+2\sqrt{2}
\end{bmatrix}
\begin{Bmatrix} \bar{F}_{x2} \\ \bar{F}_{y2} \end{Bmatrix}
$$

It should be noted that the part of the original matrix that had to be inverted could have been obtained directly from $[\bar{K}]$ simply by deleting rows and columns corresponding to zero displacements.

The values of $\bar{u}_2$ and $\bar{v}_2$ are now used to determine the reactions $\overline{F}_{x1}, \overline{F}_{y1}, \overline{F}_{x3}$ and $\overline{F}_{y3}$.

$$
\begin{Bmatrix} \overline{F}_{x1} \\ \overline{F}_{y1} \\ \overline{F}_{x3} \\ \overline{F}_{y3} \end{Bmatrix} = \begin{bmatrix} -1 & 0 \\ 0 & 0 \\ \dfrac{-1}{2\sqrt{2}} & \dfrac{1}{2\sqrt{2}} \\ \dfrac{1}{2\sqrt{2}} & \dfrac{-1}{2\sqrt{2}} \end{bmatrix} \begin{bmatrix} 1 & 1 \\ 1 & 1+2\sqrt{2} \end{bmatrix} \begin{Bmatrix} \overline{F}_{x2} \\ \overline{F}_{y2} \end{Bmatrix}
$$

$$
= \begin{bmatrix} -1 & -1 \\ 0 & 0 \\ 0 & -1 \\ 0 & -1 \end{bmatrix} \begin{Bmatrix} \overline{F}_{x2} \\ \overline{F}_{y2} \end{Bmatrix} = \begin{Bmatrix} (-\overline{F}_{x2} - \overline{F}_{y2}) \\ 0 \\ -\overline{F}_{y2} \\ -\overline{F}_{y2} \end{Bmatrix}
$$

Finally the forces in the members are calculated, using equation 2.17, as

$$
P_3 = \frac{AE}{L\sqrt{2}} \begin{bmatrix} \dfrac{-1}{\sqrt{2}} & \dfrac{1}{\sqrt{2}} \end{bmatrix} \begin{Bmatrix} (u_3 - u_2) \\ (v_3 - v_2) \end{Bmatrix} = -\sqrt{2}\,F_{y2}
$$

$$
P_1 = \frac{AE}{L} \begin{bmatrix} 1 & 0 \end{bmatrix} \begin{Bmatrix} (u_2 - u_1) \\ (v_2 - v_1) \end{Bmatrix} = F_{x2} + (1 + 2\sqrt{2})F_{y2}
$$

$$
P_2 = \frac{AE}{L} \begin{bmatrix} 0 & 1 \end{bmatrix} \begin{Bmatrix} (u_3 - u_1) \\ (v_3 - v_1) \end{Bmatrix} = 0
$$

For example, if $\overline{F}_{y2} = 0$ and $\overline{F}_{x2} = 10$ kN, $P_3 = 0$, $P_1 = 10$ kN and $P_2 = 0$.

The reader is now strongly recommended to use the matrix methods presented so far to solve a number of statically-determinate problems, since by so doing he will become familiar with the assembly process.

Numbering of nodes

The $[\overline{K}^e]$ matrix, written in global co-ordinates, for a linear member connecting nodes m and n is

$$\begin{Bmatrix} \overline{F}_{xm} \\ \overline{F}_{ym} \\ \overline{F}_{xn} \\ \overline{F}_{yn} \end{Bmatrix} = \frac{AE}{L} \begin{bmatrix} \lambda^2 & \lambda\mu & -\lambda^2 & -\lambda\mu \\ \lambda\mu & \mu^2 & -\lambda\mu & -\mu^2 \\ -\lambda^2 & -\lambda\mu & \lambda^2 & \lambda\mu \\ -\lambda\mu & -\mu^2 & \lambda\mu & \mu^2 \end{bmatrix} \begin{Bmatrix} \overline{u}_m \\ \overline{v}_m \\ \overline{u}_n \\ \overline{v}_n \end{Bmatrix}$$

From the examples worked so far, it is clear that there will be terms on the main diagonal at locations mm and nn and terms in the off-diagonal positions mn and nm. In order to keep the 'band width' of the overall matrix as small as possible, the nodes should always be numbered in such a way that the maximum difference in the node numbers is as small as possible. The following example illustrates this point.

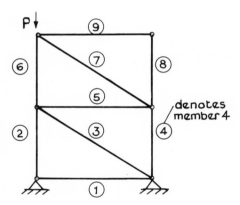

Fig. 2.13.

Consider the pin-jointed structure shown in Figure 2.13, loaded by a single load P and supported on two non-yielding bases. In Figures 2.14 and 2.15 the overall stiffness matrix $[\overline{K}^e]$ has been assembled for two different numbering schemes. Since actual computer programs normally only store the terms within the diagonal band, the benefits of numbering the joints so that the maximum difference between any two element node numbers is kept to a

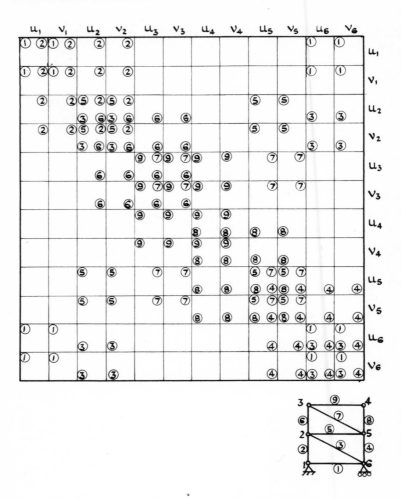

Fig. 2.14. Distribution of terms in overall stiffness matrix, for chosen node numbering: system 1

minimum is evident. The numbering used in Figure 2.15 leads to a narrow, dense, diagonal band. The rule to achieve this arrangement is simple. Numbering should commence along the narrow width of the structure. When the end of that line of nodes is reached, the node above the first numbered node is returned to and the process repeated, as illustrated in Figure 2.16. The greatest difference in node numbers is four and therefore the semi-band width of the

Columns: u_1 v_1 u_2 v_2 u_3 v_3 u_4 v_4 u_5 v_5 u_6 v_6

Rows: u_1, v_1, u_2, v_2, u_3, v_3, u_4, v_4, u_5, v_5, u_6, v_6

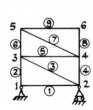

Fig. 2.15. Distribution of terms in overall stiffness matrix for chosen node numbering: system 2

resulting matrix (which includes the diagonal term) (Figure 2.16) is

number of degrees of freedom per node ×
$\qquad$ (largest difference in node numbers + 1).

In the present case, this gives $2 \times (4+1) = 10$.
$\quad$ Checking against Figures 2.14 and 2.15

Semi-band width in Figure 2.14 $= 2(5+1) = 12$.
Semi-band width in Figure 2.15 $= 2(2+1) = 6$.

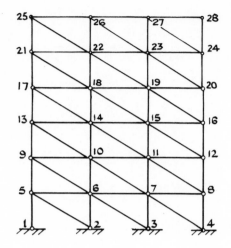

Fig. 2.16. Numbering of nodes to result in a well-banded matrix

Extension to rigid-jointed frameworks

Whereas the basic element for the pin-jointed framework was a simple pin-ended bar, the basic element for analysing rigid-jointed frameworks is the beam element shown in Figure 2.17. The beam is loaded by forces and moments at each node and is assumed to be of uniform flexural rigidity EI.

The element stiffness matrix $[K^e]$ may again be obtained by following the procedure used previously in this chapter, i.e. by imposing each of the nodal displacements in turn, determining the nodal forces set up, and then superimposing the results obtained

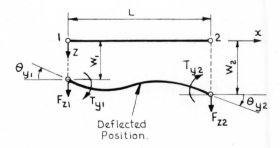

Fig. 2.17.

from each individual case. However, such a procedure is unnecessary in this case since the required relationships between the nodal forces and displacements may be obtained directly from the well-known slope–deflection equations. For the element shown in Figure 2.17, the slope–deflection equations may be written as follows:

$$T_{y1} = \left(\frac{6EI}{L^2}\right) w_1 + \left(\frac{4EI}{L}\right) \theta_{y1} - \left(\frac{6EI}{L^2}\right) w_2 + \left(\frac{2EI}{L}\right) \theta_{y2}$$

and

$$T_{y2} = \left(\frac{6EI}{L^2}\right) w_1 + \left(\frac{2EI}{L}\right) \theta_{y1} - \left(\frac{6EI}{L^2}\right) w_2 + \left(\frac{4EI}{L}\right) \theta_{y2}$$

From statical equilibrium

$$F_{z1} = -F_{z2} = \left(\frac{T_{y1} + T_{y2}}{L}\right)$$

$$= \left(\frac{12EI}{L^3}\right) w_1 + \left(\frac{6EI}{L^2}\right) \theta_{y1} - \left(\frac{12EI}{L^3}\right) w_2 + \left(\frac{6EI}{L^2}\right) \theta_{y2}$$

These equations may be written in matrix form as follows.

$$\begin{Bmatrix} F_{z1} \\ T_{y1} \\ F_{z2} \\ T_{y2} \end{Bmatrix} = \frac{EI}{L^3} \begin{bmatrix} 12 & 6L & -12 & 6L \\ 6L & 4L^2 & -6L & 2L^2 \\ -12 & -6L & 12 & -6L \\ 6L & 2L^2 & -6L & 4L^2 \end{bmatrix} \begin{Bmatrix} w_1 \\ \theta_{y1} \\ w_2 \\ \theta_{y2} \end{Bmatrix} \quad (2.18)$$

and may be summarised as

$$\{F^e\} = [K^e]\{\delta^e\}$$

thus defining the stiffness matrix $[K^e]$ for the beam element shown in Figure 2.17.

For an arbitrarily orientated beam element it is again necessary to expand $[K^e]$ to allow for displacements transforming into both u

and v displacements in the global system. The moments are unaffected. Consequently $[T]$ takes the form

$$[T] = \begin{bmatrix} \lambda & \mu & 0 & 0 & 0 & 0 \\ -\mu & \lambda & 0 & 0 & 0 & 0 \\ 0 & 0 & 1 & 0 & 0 & 0 \\ 0 & 0 & 0 & \lambda & \mu & 0 \\ 0 & 0 & 0 & -\mu & \lambda & 0 \\ 0 & 0 & 0 & 0 & 0 & 1 \end{bmatrix}$$

The analysis now proceeds along the same lines as indicated previously, the overall stiffness matrix $[K]$ being formed, the boundary conditions being applied and the resulting set of equations solved. The methods of performing these operations and, in particular, of programming them for a digital computer are discussed in later chapters.

CHAPTER THREE

Bar Finite Elements

Introduction

In the preceding chapter the basic ideas of the stiffness method of analysis were developed. In particular it was demonstrated that the sequence of operations involved in using the method to solve a problem is independent of the type of element being used. This point is very important since it means that computer programs may be written that are capable of analysing a variety of problems provided, of course, that the correct types of elements are available. The following chapters deal principally with the first stage in the stiffness analysis of a structure or component, namely the formation of the element stiffness matrix $[K^e]$.

In the preceding examples it has been possible to develop the stiffness matrix using the existing information available in standard books dealing with the strength of materials and the theory of structures and this stage has consisted merely of a recasting of these results in the appropriate form. However, for other types of finite elements, e.g. a triangular element subjected to plane stress, no such results are available. It is therefore necessary that a general procedure for the derivation of the element stiffness matrix $[K^e]$ be advanced. Since the exact nature of the final stage in the complete analysis, i.e. the evaluation of the internal element forces from a knowledge of the nodal displacements of the element, also depends upon the type of element being used the general procedure should also contain a step in which the force–displacement relationship for the element is established.

In this chapter the seven basic steps employed in the derivation of the stiffness characteristics for a finite element are presented. In order that the reader can more easily understand the process each step is first explained in general terms and then again with particular reference to the beam element, the stiffness matrix of which was given at the end of Chapter 2. The seven steps are listed, together with brief details of the operations involved, in Appendix 1. Subsequent chapters demonstrate that, by following the same basic steps, it is possible to derive the stiffness matrices of several other types of finite elements including both triangles and rectangles subjected to plane stress and rectangles subjected to bending.

Stiffness matrix for beam element

Step I: Identify problem

General case
The first step is to choose a suitable co-ordinate and node numbering system for the element.

Since the degrees of freedom that the chosen finite element possesses are known, the nodal displacement vector $\{\delta^e\}$ and the nodal load vector $\{F^e\}$ for the element can be established. The stiffness matrix $[K^e]$ for this individual element is then defined by equation I.

$$\{F^e\} = [K^e]\{\delta^e\} \tag{I}$$

Specific case of beam element
Consider a beam element with a uniform cross-section that forms part of a continuous structure. For such an element the co-ordinate and node numbering system shown in Figure 2.17 may be used, where the y-axis is normal to the plane of the paper. In this Figure the circles at the ends of the element merely indicate the position of the nodes and do not denote hinges. Since the problem is one of beam

bending, if, as in normal engineering frame analysis, axial deformations are ignored, only two displacements need be considered at each node, a deflection normal to the beam w and a rotation about the y-axis θ_y. Therefore, the total number of degrees of freedom for each beam element is four, as shown in Figure 2.17. At this stage the term 'degrees of freedom' may be regarded as meaning the same as nodal displacements. However, as is mentioned later, some more advanced elements involve terms other than simple displacements, e.g. curvatures, and in such cases a more general definition is therefore necessary.

Associated with the rotation and transverse displacement at each end of the element are the corresponding nodal forces, a moment T_y and a shearing force F_z as shown in Figure 2.17.

The vectors for the displacements and forces at node 1 may then be written respectively as

$$\{\delta_1^e\} = \begin{Bmatrix} w_1 \\ \theta_{y1} \end{Bmatrix}, \qquad \{F_1^e\} = \begin{Bmatrix} F_{z1} \\ T_{y1} \end{Bmatrix}$$

The complete vectors of nodal displacements and nodal loads for the beam 1–2 therefore take the form

$$\{\delta^e\} = \begin{Bmatrix} \{\delta_1^e\} \\ \{\delta_2^e\} \end{Bmatrix} = \begin{Bmatrix} w_1 \\ \theta_{y1} \\ w_2 \\ \theta_{y2} \end{Bmatrix}, \qquad \{F^e\} = \begin{Bmatrix} \{F_1^e\} \\ \{F_2^e\} \end{Bmatrix} = \begin{Bmatrix} F_{z1} \\ T_{y1} \\ F_{z2} \\ T_{y2} \end{Bmatrix}$$

Since each of these vectors contains four terms the element stiffness matrix $[K^e]$ is of the order 4×4.

Step II: Select suitable displacement function

General case
A displacement function that uniquely defines the state of displacement at all points within the element is chosen.

The displacement pattern can conveniently be represented by a polynomial expression and since the aim is to express the displacements at any point $\{\delta(x, y)\}$ in terms of the nodal displacements

$\{\delta^e\}$, the assumed polynomial must contain one unknown co-efficient for each degree of freedom possessed by the element. The state of displacement at any point (x, y) within the element may then be written in matrix form as shown in equation II.

$$\{\delta(x, y)\} = [f(x, y)]\{\alpha\} \tag{II}$$

where $\{\alpha\}$ is a column vector of the as yet unknown coefficients of the polynomial function $[f(x, y)]$.

Specific case of beam element
The displacement of any point within the element may be defined by the two components, the deflection w and the rotation θ_y. Thus the displacement vector is given by

$$\{\delta(x, y)\} = \begin{Bmatrix} w \\ \theta_y \end{Bmatrix}$$

Since the element possesses four degrees of freedom (w_1, θ_{y1}, w_2 and θ_{y2}), four unknown coefficients must appear in the polynomial representing the displacement pattern. Assume the displacement w is as given by equation 3.1.

$$w = \alpha_1 + \alpha_2 x + \alpha_3 x^2 + \alpha_4 x^3 \tag{3.1}$$

where α_1, α_2, α_3 and α_4 are as yet unknown coefficients. Since $\theta_y = dw/dx$, then from equation 3.1

$$\theta_y = \alpha_2 + 2\alpha_3 x + 3\alpha_4 x^2 \tag{3.1a}$$

Thus from equations 3.1 and 3.1a the displacement vector $\{\delta(x, y)\}$ in equation 3.2 can be obtained.

$$\{\delta(x, y)\} = \begin{Bmatrix} w \\ \theta_y \end{Bmatrix} = \begin{bmatrix} 1 & x & x^2 & x^3 \\ 0 & 1 & 2x & 3x^2 \end{bmatrix} \begin{Bmatrix} \alpha_1 \\ \alpha_2 \\ \alpha_3 \\ \alpha_4 \end{Bmatrix} \tag{3.2}$$

which defines the matrix $[f(x, y)]$ and the vector $\{\alpha\}$ of equation II.

Step III: Relate general displacements within an element to its nodal displacements

General case
The coefficients of the displacement function $\{\alpha\}$ are now expressed in terms of the nodal displacements $\{\delta^e\}$ and hence by substituting in equation II the displacements at any point within the element are related to the nodal displacements $\{\delta^e\}$.

Since $\{\delta(x, y)\}$ represents the displacement at any point (x, y) the nodal displacements can be obtained from it by simply substituting the appropriate nodal co-ordinates into equation 3.2 (e.g. those at node 1, into equation II). Therefore, for node 1

$$\{\delta_1^e\} = \{\delta(x_1, y_1)\} = [f(x_1, y_1)]\{\alpha\}$$

Proceeding similarly for all other nodes, for the case of an element containing n nodes,

$$\{\delta^e\} = \begin{Bmatrix} \{\delta_1^e\} \\ \{\delta_2^e\} \\ \vdots \\ \{\delta_n^e\} \end{Bmatrix} = \begin{bmatrix} [f(x_1, y_1)] \\ [f(x_2, y_2)] \\ \vdots \\ [f(x_n, y_n)] \end{bmatrix} \{\alpha\}$$

or

$$\{\delta^e\} = [A]\{\alpha\} \tag{IIIa}$$

Since the matrix $[A]$ is now known, the vector of unknown coefficients $\{\alpha\}$ may be obtained by inverting the expression in equation IIIa to give

$$\{\alpha\} = [A]^{-1}\{\delta^e\}$$

Substituting for $\{\alpha\}$ in equation II gives the required relationship between the displacements at any point within the element $\{\delta(x, y)\}$ and the nodal displacements $\{\delta^e\}$ as

$$\{\delta(x, y)\} = [f(x, y)][A]^{-1}\{\delta^e\} \tag{III}$$

Specific case of beam element
For the simple beam element shown in Figure 2.17 the co-ordinates of the nodes are 0 and L. The chosen expression defining $\{\delta(x, y)\}$, (see equation 3.1) is

$$w = \alpha_1 + \alpha_2 x + \alpha_3 x^2 + \alpha_4 x^3$$

$$\theta_y = \frac{dw}{dx} = \alpha_2 + 2\alpha_3 x + 3\alpha_4 x^2$$

At node 1, $x=0$ and therefore

$$w_1 = \alpha_1 \quad \text{and} \quad \theta_{y1} = \alpha_2$$

At node 2, $x=L$ and therefore

$$w_2 = \alpha_1 + \alpha_2 L + \alpha_3 L^2 + \alpha_4 L^3$$
$$\theta_{y2} = \alpha_2 + 2\alpha_3 L + 3\alpha_4 L^2$$

Recasting these results in matrix form

$$\begin{Bmatrix} w_1 \\ \theta_{y1} \\ w_2 \\ \theta_{y2} \end{Bmatrix} = \begin{bmatrix} 1 & 0 & 0 & 0 \\ 0 & 1 & 0 & 0 \\ 1 & L & L^2 & L^3 \\ 0 & 1 & 2L & 3L^2 \end{bmatrix} \begin{Bmatrix} \alpha_1 \\ \alpha_2 \\ \alpha_3 \\ \alpha_4 \end{Bmatrix} \tag{3.3}$$

The matrix $[A]$ is therefore defined by the square 4×4 matrix in equation 3.3.

To obtain matrix $[A]^{-1}$ it is necessary to invert equation 3.3. Due to the form of this equation, this can conveniently be done by solving the set of simultaneous equations as shown below. Inspecting equation 3.3 immediately reveals that $\alpha_1 = w_1$ and $\alpha_2 = \theta_{y1}$. It therefore only remains to solve the last two equations given in equation 3.3 to obtain

$$\alpha_3 = \frac{3}{L^2}(-w_1 + w_2) - \frac{1}{L}(2\theta_{y1} + \theta_{y2})$$

$$\alpha_4 = \frac{2}{L^3}(w_1 - w_2) + \frac{1}{L^2}(\theta_{y1} + \theta_{y2})$$

Rewriting these results in matrix form yields

$$\begin{Bmatrix} \alpha_1 \\ \alpha_2 \\ \alpha_3 \\ \alpha_4 \end{Bmatrix} = \begin{bmatrix} 1 & 0 & 0 & 0 \\ 0 & 1 & 0 & 0 \\ -\dfrac{3}{L^2} & -\dfrac{2}{L} & \dfrac{3}{L} & -\dfrac{1}{L} \\ \dfrac{2}{L^3} & \dfrac{1}{L^2} & -\dfrac{2}{L^3} & \dfrac{1}{L^2} \end{bmatrix} \begin{Bmatrix} w_1 \\ \theta_{y1} \\ w_2 \\ \theta_{y2} \end{Bmatrix} \qquad (3.4)$$

The matrix $[A]^{-1}$ corresponds to the square 4×4 matrix in equation 3.4.

Step IV: Strain–displacement relationships

General case
The strains $\varepsilon(x, y)$ at any point (x, y) within the element are now related to the displacements $\delta(x, y)$ at that point and hence to the nodal displacements $\{\delta^e\}$.

The strains at any point within an element may be obtained from the chosen displacement function by differentiation, the exact form of the differentiation depending upon the type of problem being considered. For example, for a plane elasticity problem the strains correspond to the first derivative of displacements, while for flexural problems the strains are associated with the curvature of the element and correspond to the second derivative of displacements. In general

$$\{\varepsilon(x, y)\} = \{\text{differential of } \delta(x, y)\}$$

The exact form of this expression for any particular class of problem may be obtained from the theory of elasticity. Using the expression for $\{\delta(x, y)\}$ given by equation III and noting that both $[A]^{-1}$ and $\{\delta^e\}$ are independent of x and y, the strain vector $\{\varepsilon(x, y)\}$ is given by

$$\{\varepsilon(x, y)\} = [\text{differential of } f(x, y)][A]^{-1}\{\delta^e\}$$

Defining the matrix $[\text{differential of } f(x, y)]$ as being equal to a matrix $[C]$ the above equation can be written as

$$\{\varepsilon(x, y)\} = [C][A]^{-1}\{\delta^e\}$$

where, in general, $[C]$ contains terms in x and y.

This is the required relationship between the strains at any point within the element and the nodal displacements. Let

$$[C][A]^{-1} = [B]$$

Then the relationship between the strains at any point and the nodal displacement is given by equation IV.

$$\{\varepsilon(x, y)\} = [B]\{\delta^e\} \tag{IV}$$

Specific case of beam element
For the particular problem of a beam element the only 'strain' that need be considered is the curvature about the y-axis. Therefore from equation 3.1, the strain vector is given by

$$\{\varepsilon(x, y)\} = \frac{-d^2w}{dx^2} = -2\alpha_3 - 6\alpha_4 x$$

Rewriting this result in matrix form gives

$$\{\varepsilon(x, y)\} = \begin{bmatrix} 0 & 0 & -2 & -6x \end{bmatrix} \begin{Bmatrix} \alpha_1 \\ \alpha_2 \\ \alpha_3 \\ \alpha_4 \end{Bmatrix} \tag{3.5}$$

During the course of step III it was established that $\{\alpha\} = [A]^{-1}\{\delta^e\}$. It is thus possible to substitute for $\{\alpha\}$ in equation 3.5 to obtain

$$\{\varepsilon(x, y)\} = \begin{bmatrix} 0 & 0 & -2 & -6x \end{bmatrix} \begin{bmatrix} 1 & 0 & 0 & 0 \\ 0 & 1 & 0 & 0 \\ -\dfrac{3}{L^2} & -\dfrac{2}{L} & \dfrac{3}{L^2} & -\dfrac{1}{L} \\ \dfrac{2}{L^3} & \dfrac{1}{L^2} & -\dfrac{2}{L^3} & \dfrac{1}{L^2} \end{bmatrix} \begin{Bmatrix} w_1 \\ \theta_{y1} \\ w_2 \\ \theta_{y2} \end{Bmatrix}$$

After matrix multiplication the above equation reduces to

$$\{\varepsilon(x, y)\} = \begin{bmatrix} \dfrac{6}{L^2} - \dfrac{12x}{L^3} & \dfrac{4}{L} - \dfrac{6x}{L^2} & -\dfrac{6}{L^2} + \dfrac{12x}{L^3} & \dfrac{2}{L} - \dfrac{6x}{L^2} \end{bmatrix} \begin{Bmatrix} w_1 \\ \theta_{y1} \\ w_2 \\ \theta_{y2} \end{Bmatrix}$$

Thus matrix $[B]$ of equation IV has been obtained.

Step V: Stress–strain relationships

General case
The internal stresses $\{\sigma(x, y)\}$ occurring in the element are now related to the strains $\{\varepsilon(x, y)\}$.

Since a relationship between the internal strains and the nodal displacements $\{\delta^e\}$ is already known, the internal stresses $\{\sigma(x, y)\}$ can be related to the nodal displacements. Clearly in this step the elastic properties of the element have to be taken into consideration. In general,

$$\{\sigma(x, y)\} = [D]\{\varepsilon(x, y)\}$$

where $[D]$ is termed the elasticity matrix and contains the elastic properties, i.e. quantities such as the modulus of elasticity E and Poisson's ratio v. Since from equation IV it is already known that $\{\varepsilon(x, y)\} = [B]\{\delta^e\}$,

$$\{\sigma(x, y)\} = [D][B]\{\delta^e\} \qquad \text{(V)}$$

Specific case of beam element
For the beam element the 'stress' $\{\sigma(x, y)\}$ and 'strain' $\{\varepsilon(x, y)\}$ correspond to the internal moment in the beam M_y and the curvature $-d^2w/dx^2$. Thus

$$M_y = -EI\,\frac{d^2w}{dx^2}$$

(It may be noted that this relationship is obtained from the simple theory of bending, namely $M/I = f/y = E/R$, where the curvature $1/R$ may be approximated by $-d^2w/dx^2$).

Therefore in this particular case matrix $[D]$ contains only a single term corresponding to the flexural rigidity EI. In general, however, $[D]$ is of a higher order. Equation V therefore becomes

$$\{\sigma(x, y)\} = [EI]\left[\frac{6}{L^2} - \frac{12x}{L^3} \quad \frac{4}{L} - \frac{6x}{L^2} \quad -\frac{6}{L^2} + \frac{12x}{L^3} \quad \frac{2}{L} - \frac{6x}{L^2}\right]\begin{Bmatrix} w_1 \\ \theta_{y1} \\ w_2 \\ \theta_{y2} \end{Bmatrix}$$

Step VI: Relate nodal loads to nodal displacements

General case

The internal stresses $\{\sigma(x, y)\}$ are now replaced by statically equivalent nodal loads $\{F^e\}$ and hence the nodal loads are related to the nodal displacements $\{\delta^e\}$ thereby defining the required element stiffness matrix $[K^e]$.

The principle of virtual work is used to determine the set of nodal loads that is statically equivalent to the internal stresses. The condition of equivalence may be expressed as follows: during any virtual displacement imposed on the element, the total external work done by the nodal loads must equal the total internal work done by the stresses. An arbitrary set of nodal displacements represented by the vector $\{\delta^{*e}\}$ is selected where

$$\{\delta^{*e}\} = \begin{Bmatrix} \{\delta_1{}^{*e}\} \\ \{\delta_2{}^{*e}\} \\ \vdots \\ \{\delta_n{}^{*e}\} \end{Bmatrix}$$

The external work done by the nodal loads W_{ext} is given by

$$W_{ext} = \{\delta_1{}^{*e}\}\{F_1{}^e\} + \{\delta_2{}^{*e}\}\{F_2{}^e\} + \cdots + \{\delta_n{}^{*e}\}\{F_n{}^e\}$$
$$= \{\delta^{*e}\}^T\{F^e\}$$

If the arbitrarily imposed displacements cause strains $\{\varepsilon(x, y)^*\}$ at a point within the element where the actual stresses are $\{\sigma(x, y)\}$, then the internal work done per unit volume is given by

$$W_{int} = \{\varepsilon(x, y)^*\}^T\{\sigma(x, y)\}$$

and the total internal work is obtained by integrating over the volume of the element, namely

$$\int^v W_{int} \, d\,(\text{vol}) = \int^v \{\varepsilon(x, y)^*\}^T\{\sigma(x, y)\} \, d(\text{vol})$$

Now from equation IV it is known that the strains set up at any point in the element are given in terms of the nodal displacements by $\{\varepsilon(x, y)\} = [B]\{\delta^e\}$. Hence when nodal displacements $\{\delta^{*e}\}$ are imposed, the corresponding strains may be written as

$$\{\varepsilon(x, y)^*\} = [B]\{\delta^{*e}\}$$

Furthermore, from equation V the actual stresses in the element are known to be related to the actual nodal displacements as

$$\{\sigma(x, y)\} = [D][B]\{\delta^e\}$$

Therefore, these expressions may be substituted into the virtual-work equation for the internal work to obtain

$$\int^v W_{int}\, d(\text{vol}) = \int^v [B]^T\{\delta^{*e}\}[D][B]\{\delta^e\}\, d(\text{vol})$$

and

$$W_{ext} = \{\delta^{*e}\}^T\{F^e\}$$

The final operation is to equate internal and external work done during the system of virtual displacements $\{\delta^{*e}\}$. Since the basic principle of virtual displacements is valid for any system of applied displacements the selection of the system of virtual nodal displacements may be chosen at will. For present purposes, it is convenient to assume that unit values of the nodal displacements are applied. Then equating the internal and external work gives

$$\{F^e\} = [\int^v [B]^T[D][B]\, d(\text{vol})]\{\delta^e\} \qquad (\text{VI})$$

On comparing equation VI with equation I, which is restated below,

$$\{F^e\} = [K^e]\{\delta^e\}$$

it is clear that the required element stiffness matrix $[K^e]$ is given by the expression in the large square brackets in equation VI. Therefore

$$[K^e] = \int^v [B]^T[D][B]\, d(\text{vol})$$

Therefore, to evaluate the element stiffness matrix in the general case, it is necessary to formulate matrix $[B]$ as set out in step IV from matrices $[A]^{-1}$ and $[C]$, and matrix $[D]$ as set out in step V, and then to perform the matrix multiplications and integrations defined by equation VI.

Specific case of beam element
Equation VI has been established for a completely general case where, in order to obtain the total internal work done during a virtual displacement of the element, the product of the internal stresses and strains has to be integrated over the total volume of the element. For

the particular case of a beam element, the internal 'stress' $\{\sigma(x, y)\}$ corresponds to the internal moment per unit length M_y so that, in order to evaluate the total internal work done, the product of the internal moment per unit length and its associated curvature must be integrated over the total length of the element. The expression $\int^v d(\text{vol})$ in equation VI for the general case must thus be replaced by the expression $\int_0^L dx$ for the specific case of a beam element, so that equation VI may be rewritten as

$$\{F^e\} = [\int_0^L [B]^T[D][B]dx]\{\delta^e\}$$

All the matrices in this equation have already been worked out explicitly. The matrix $[K^e]$, which is given by the expression in the outer square brackets in equation VI, may therefore be obtained from:

$$[K^e] = \int_0^L \begin{bmatrix} \dfrac{6}{L^2} - \dfrac{12x}{L^3} \\[2mm] \dfrac{4}{L} - \dfrac{6x}{L^2} \\[2mm] -\dfrac{6}{L^2} + \dfrac{12x}{L^3} \\[2mm] \dfrac{2}{L} - \dfrac{6x}{L^2} \end{bmatrix} [EI] \begin{bmatrix} \dfrac{6}{L^2} - \dfrac{12x}{L^3} & \dfrac{4}{L} - \dfrac{6x}{L^2} & -\dfrac{6}{L^2} + \dfrac{12x}{L^3} & \dfrac{2}{L} - \dfrac{6x}{L^2} \end{bmatrix} d$$

Performing the multiplication gives the expression shown below as equation 3.6 (p. 47) and then performing the integration yields finally the 4×4 matrix for $[K^e]$ given by equation 3.7 (p. 47).

Step VII: Stress–displacement relationships

General case
Finally the stress–displacement matrix $[H]$ that relates the internal stresses in the element $\{\sigma(x, y)\}$ to its nodal displacements $\{\delta^e\}$ is determined. This then enables the internal stresses to be evaluated once the overall problem has been solved for the nodal displacements.

$$[K^e] = [EI] \int_0^L \begin{bmatrix} \frac{36}{L^4} - \frac{144x}{L^5} + \frac{144x^2}{L^6} & \frac{24}{L^3} - \frac{84x}{L^4} + \frac{72x^2}{L^5} & -\frac{36}{L^4} + \frac{144x}{L^5} - \frac{144x^2}{L^6} & \frac{12}{L^3} - \frac{60x}{L^4} + \frac{72x^2}{L^5} \\[2mm] \frac{24}{L^3} - \frac{84x}{L^4} + \frac{72x^2}{L^5} & \frac{16}{L^2} - \frac{48x}{L^3} + \frac{36x^2}{L^4} & -\frac{24}{L^3} + \frac{84x}{L^4} - \frac{72x^2}{L^5} & \frac{8}{L^2} - \frac{36x}{L^3} + \frac{36x^2}{L^4} \\[2mm] -\frac{36}{L^4} + \frac{144x}{L^5} - \frac{144x^2}{L^6} & -\frac{24}{L^3} + \frac{84x}{L^4} - \frac{72x^2}{L^5} & \frac{36}{L^4} - \frac{144x}{L^5} + \frac{144x^2}{L^6} & -\frac{12}{L^3} + \frac{60x}{L^4} - \frac{72x^2}{L^5} \\[2mm] \frac{12}{L^3} - \frac{60x}{L^4} + \frac{72x^2}{L^5} & \frac{8}{L^2} - \frac{36x}{L^3} + \frac{36x^2}{L^4} & -\frac{12}{L^3} + \frac{60x}{L^4} - \frac{72x^2}{L^5} & \frac{4}{L^2} - \frac{24x}{L^3} + \frac{36x^2}{L^4} \end{bmatrix} dx$$

$$\tag{3.6}$$

$$[K^e] = \frac{EI}{L^3} \begin{bmatrix} 12 & 6L & -12 & 6L \\ 6L & 4L^2 & -6L & 2L^2 \\ -12 & -6L & 12 & -6L \\ 6L & 2L^2 & -6L & 4L^2 \end{bmatrix}$$

$$\tag{3.7}$$

It should be noted that this matrix corresponds to equation 2.18.

The required relationship has in fact already been developed in step V and is defined by equation V as

$$\{\sigma(x, y)\} = [D][B]\{\delta^e\}$$

The stress–displacement matrix $[H]$ is therefore given by the product of the $[D]$ and $[B]$ matrices, i.e.

$$[H] = [D][B] \qquad\qquad\qquad\text{(VII)}$$

In general, matrix $[H]$ contains terms in x and y and therefore relates the stress at any point (x, y) within the element to the nodal displacements $\{\delta^e\}$. It is sometimes convenient to evaluate stresses at the nodal points and this requires the use of a matrix $[H^e]$, where $[H^e]$ is obtained from $[H]$ simply by substituting the various nodal co-ordinates.

Specific case of beam element
Taking the expression for $[D]$ and $[B]$ defined earlier for the beam element gives $[H]$ for this element as

$$[H] = EI \left[\frac{6}{L^2} - \frac{12x}{L^3} \quad \frac{4}{L} - \frac{6x}{L^2} \quad \frac{-6}{L^2} + \frac{12x}{L^3} \quad \frac{2}{L} - \frac{6x}{L^2} \right]$$

Since the co-ordinates of the element being considered are 0 and L the matrix $[H^e]$ becomes

$$[H^e] = \frac{EI}{L^2} \begin{bmatrix} 6 & 4L & -6 & 2L \\ -6 & -2L & 6 & -4L \end{bmatrix}$$

Thus both the stiffness matrix $[K^e]$ and the stress–displacement matrix $[H]$ have now been obtained for the beam element and at the same time the general procedure, which involves seven basic steps, for developing these matrices for any type of element has been established. Subsequent chapters show how, by following the same basic procedure, $[K^e]$ and $[H]$ may be determined for other types of elements.

It should be noted that, in this particular case, because the chosen displacement function corresponds to the correct displaced shape of the element in simple flexure, the matrix $[K^e]$ obtained from the present approach is identical to that given in equation 2.18 which was obtained from the slope–deflection equations. In general the chosen displacement functions will only approximate to the correct displaced shape of the element.

CHAPTER FOUR

Finite Elements of Continua

Introduction

The finite element method is an extension of the analysis techniques of ordinary framed structures to two- and three-dimensional structures such as plates and shells. The method was pioneered in the aircraft industry where there was an urgent need for accurate analysis of complex airframes. The availability of automatic digital computers from 1950 onwards contributed to the rapid development of matrix methods during this period.

The idealisation of elastic continua by one-dimensional elements had been undertaken for example by Hrennikoff[1]* in 1941 and McHenry[2] in 1943 for plane elasticity problems using their lattice analogy, for plates using grillage systems by Newmark[3] and for shells using a space truss analogy by Parikh and Norris[4]. The formulation of the matrix method of structural analysis, in particular the contributions of Argyris[5], was the basis from which the finite element was developed by Clough[6] and his co-workers. The basic concept of the finite element method, as with the general matrix structural analysis of frames, is that the structure can be considered to be an assemblage of individual structural elements.

* Superscript figures refer to references at end of chapter.

The new idea in the finite element method is the use of two- and three-dimensional elements for the idealisation of a continuum. The paper in 1956 by Turner, Clough, Martin and Topp[6] in which a continuous structure, part of an aircraft, was analysed as an assemblage of two-dimensional elements may be regarded as the starting point of the finite element method.

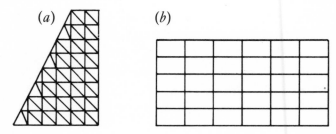

Fig. 4.1. Typical finite element idealisations. *a* Triangular elements. *b* Rectangular elements

In the previous chapters it has been shown that, in a matrix analysis of a structure, the structure is divided into a number of elements interconnected at a discrete number of nodal points. Since skeletal structures consist of linear members connected together at a number of joints the subdivision required by the matrix method is a natural one and such structures can be readily analysed using matrix methods. In a continuum structure such as a flat plate, a corresponding natural subdivision does not exist so that the continuum has to be artificially divided into a number of elements before the matrix method of analysis can be applied. The artificial elements, which are termed 'finite elements' or discrete elements, are usually chosen to be either rectangular or triangular in shape. Figure 4.1*a* shows how a retaining wall can be divided into triangular elements and Figure 4.1*b* gives an example of a rectangular plate divided into rectangular elements.

In reality, these elements are connected together along their common boundaries. However, in order to make a solution by the matrix method of structural analysis possible, it is assumed in the case of simple elements that these elements are only interconnected at their nodes. This assumption by itself means that continuity requirements are only satisfied at the nodal points. Clearly the relaxation of continuity requirements along the sides of the elements would make the structure very much more flexible than it actually is since it would allow the type of action shown in Figure 4.2 to develop. However, in the finite element method, the individual elements are constrained to deform in specific patterns. Hence, although continuity is only specified at the nodal points, the choice of a suitable pattern of deflection for the elements can lead to the satisfaction of some, if not all, of the continuity requirements along the sides of adjacent elements. Hence, as stated by Clough, 'finite elements are not merely pieces cut from the original structure, but are special types of elastic elements constrained to deform in specific patterns such that the overall continuity of the assemblage tends to be maintained'.

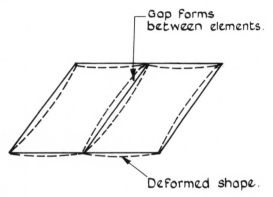

Fig. 4.2. Type of deformation that could occur if nodal continuity only is enforced

The analysis of a continuum thus differs from the analysis of a skeletal structure in two basic aspects only, namely in the initial subdivision into elements and in the derivation of the element stiffness characteristics. These two aspects are now discussed further.

Subdivision of structure

The nature of the finite element idealisation indicates that, in general, the accuracy of the solution increases with the number of elements taken. However, it must also be realised that as the number of elements taken increases, the computer time required to obtain a solution also increases, with a consequent increase in cost.

In some solutions a graded division into elements may be used to allow a more detailed study of regions within the structure where high concentrations of stress are expected, e.g. around openings and close to concentrated loads, as shown in Figure 4.3. Such a selective distribution of elements is efficient and can lead to economy in solution time without any loss of accuracy. It is impossible to generalise as to the number of elements required to give satisfactory solutions since this depends on the particular problem being considered. The choice of a suitable subdivision for a particular structure should, if possible, be based on previous reported experience of similar solutions. If this is not feasible, then a number of solutions involving different mesh sizes must be carried out to test the rate of convergence of the solution for the particular problem. Typical examples are given later in the text.

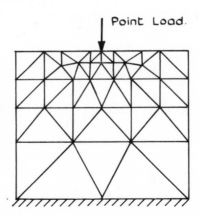

Fig. 4.3. Graded mesh

The system of external loads acting on the actual structure has to be replaced by an equivalent system of forces concentrated at the element nodes. In the case of concentrated loads, the finite element mesh would clearly be chosen to ensure that a node occurs at the point of application of the load. For distributed loading, equivalent nodal point loads must be calculated. This can be done by using the principle of virtual work and equating the work done by the actual distributed loads to the work done by the equivalent nodal forces during a virtual displacement. However, it is usually possible by inspection to allocate statically equivalent nodal forces.

Consider for example the beam shown in Figure 4.4, which has a span of 10 m and is carrying a uniformly distributed load of 10 kN/m. Then for the finite element idealisation shown, the equivalent nodal forces are as indicated. These applied nodal loads form the load vector $\{F\}$ for the overall structure and are not to be confused with the nodal force vector for an individual element $\{F^e\}$ which is statically equivalent to the internal stresses in the element. At any node the external applied load, i.e. a term from within the $\{F\}$ vector, must be in equilibrium with the sum of the nodal forces, i.e. terms from the $\{F^e\}$ vectors, of the elements meeting at that node. When no external load is applied at a particular node, then the element nodal forces at that node are in equilibrium.

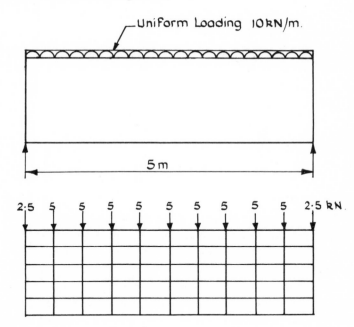

Fig. 4.4. Nodal forces representing uniformly distributed load

There are many different types of finite elements available to the engineer, each type having its own particular advantage. Some of these, together with their applications, are discussed in later chapters. The two types most commonly used are the rectangularly-shaped and triangularly-shaped elements. These are adequate for the analysis of many plane elasticity, plate flexure and shell problems, the triangular element being more versatile than the rectangular one since it can be used for the analysis of plates of various shapes and lends itself more readily to mesh grading.

Derivation of element stiffness characteristics

The procedure for deriving the stiffness characteristics of finite elements may be split up into the seven basic steps which have already been discussed for the simple beam element. In the following chapters each of these steps is discussed and a detailed derivation for a number of elements presented.

It is realised that the finite element displacement method represents an application of the variational principle of minimum potential energy. This has encouraged researchers to seek to use variational methods in the development of the finite element method. However, further consideration of the variational origins of the finite element method is left until Chapter 12.

References

1. HRENNIKOFF, A. Solution of problems in elasticity by the framework method. *Journal of Applied Mechanics*, Vol. 8, 1941, pp. A169–A175.
2. McHENRY, D. A lattice analogy for the solution of plane stress problems. *Proceedings of the Institution of Civil Engineers*, Vol. 21, No. 2, 1943–44, pp. 59–82.
3. NEWMARK, N. M. Numerical methods of analysis in bars, plates and elastic bodies. In *Numerical methods of analysis in engineering*, edited by L. E. Grinter. New York, The Macmillan Co., 1949.
4. PARIKH, K. S. and NORRIS, C. H. Analysis of shells using framework analogy. *Proceedings of the World Conference on Shell Structures, San Francisco*, 1962.
5. ARGYRIS, J. H. *Energy theorems and structural analysis*. London, Butterworth (Publishers) Ltd, 1960.
6. TURNER, M. J., CLOUGH, R. W., MARTIN, H. C. and TOPP, L. J. Stiffness and deflection analysis of complex structures. *Journal of Aeronautical Science*, Vol. 23, No. 9, 1956, pp. 805–823.

CHAPTER FIVE

Triangular Finite Element for Plane Elasticity

Introduction

Plane elasticity problems involve continua that are loaded in their plane. Where the continuum is loaded by forces normal to the plane out-of-plane displacements (e.g. bending of a plate) are induced, and such problems are considered in Chapters 7, 8, 9 and 11.

Plane elasticity problems may be separated into two separate classes, namely plane stress problems and plane strain problems[1]. In a plane stress problem, the continuum (such as a plate) is thin relative to other dimensions, and stresses normal to the plane are neglected. Problems in this category include diaphragm plates in box sections and plate girder webs where the applied loads are in the plane of the member. Figure 5.1 shows a perforated tension strip and the web of a deep beam, both being examples of plane stress.

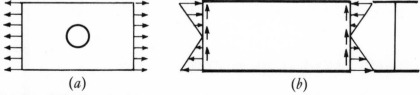

Fig. 5.1. Examples of members in plane stress. *a* Perforated tension strip. *b* Plate girder web

In a plane strain problem, the strain normal to the plane of loading is assumed to be zero. Thus the stresses and displacements in a retaining wall can be analysed by taking a transverse slice out of the wall and assuming zero strains normal to the plane of the slice. Figure 5.2 shows such an example of plane strain.

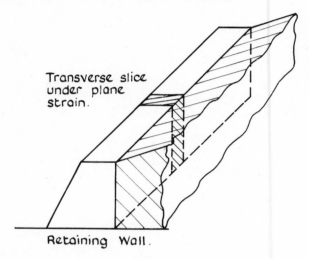

Fig. 5.2. Example of plane strain conditions

It has been shown in Chapter 4 that the essential difference between the analysis of a skeletal structure and a continuous structure by matrix methods is that the continuum must be 'idealised' as a number of finite elements (or discrete elements) interconnected only at their nodes. The displacements of these nodes are the basic unknown parameters of the problem, in just the same manner as are the unknown joint displacements in the analysis of a skeletal structure. The first task is to 'separate' the continuum into finite elements by making imaginary cuts and in this chapter one of the simplest yet most versatile shapes, namely the triangle, is considered.

Derivation of triangular element stiffness matrix

The basic steps in the derivation of the element stiffness matrix for a triangular element in plane stress and plane strain are now presented.

These basic steps are given in Appendix 1 and are identical to those used earlier for the uniform beam element. Readers are strongly recommended fully to acquaint themselves with this chart and to refer to it frequently when working through the following chapters.

Step I: Choose suitable co-ordinate system and number nodes

The Cartesian co-ordinate system shown in Figure 5.3 is used and the three nodes numbered 1, 2, 3 using an anticlockwise convention. The positions of these nodes in terms of the Cartesian co-ordinates are (x_1, y_1), (x_2, y_2) and (x_3, y_3).

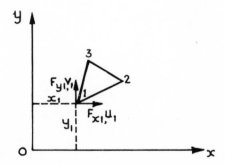

Fig. 5.3. Co-ordinates and node numbering for plane elasticity triangular element

For a plane elasticity problem, where all displacements are in the plane, the element has two degrees of freedom at each node, as shown in Figure 5.4, making a total of six degrees of freedom (u_1, v_1, u_2, v_2, u_3, v_3) for the triangular element. The corresponding forces are (F_{x1}, F_{y1}, F_{x2}, F_{y2}, F_{x3}, F_{y3}).

Using matrix notation, the displacement vector at node 1 may be written as

$$\{\delta_1\} = \begin{Bmatrix} u_1 \\ v_1 \end{Bmatrix}$$

and the corresponding force vector at node 1 may be written

$$\{F_1\} = \begin{Bmatrix} F_{x1} \\ F_{y1} \end{Bmatrix}$$

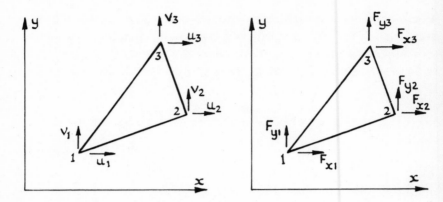

Fig. 5.4. Nodal displacements and forces

The complete displacement and force vectors for the triangular element may be written as

$$\{\delta^e\} = \begin{Bmatrix} \{\delta_1\} \\ \{\delta_2\} \\ \{\delta_3\} \end{Bmatrix} = \begin{Bmatrix} u_1 \\ v_1 \\ u_2 \\ v_2 \\ u_3 \\ v_3 \end{Bmatrix} \tag{5.1}$$

$$\{F^e\} = \begin{Bmatrix} \{F_1\} \\ \{F_2\} \\ \{F_3\} \end{Bmatrix} = \begin{Bmatrix} F_{x1} \\ F_{y1} \\ F_{x2} \\ F_{y2} \\ F_{x3} \\ F_{y3} \end{Bmatrix} \tag{5.2}$$

Since each of these vectors contains six terms, the element stiffness matrix $[K^e]$ is a 6×6 matrix for this plane elasticity triangle.

$$\{F^e\} = [K^e]\{\delta^e\} \tag{I}$$

Step II: Choose displacement function $[f(x, y)]$ that defines displacement $\{\delta(x, y)\}$ at any point in element

For a plane elasticity problem, the displacement of any point can be obtained by considering the u and v movements in the x and y directions respectively. Since there are six degrees of freedom in this case, six unknown coefficients $(\alpha_1, \alpha_2 \ldots \alpha_6)$ are required in the polynomial representing the permitted displacement pattern. The simplest representation is given by the two linear polynomials

$$\left.\begin{aligned} u &= \alpha_1 + \alpha_2 x + \alpha_3 y \\ v &= \alpha_4 + \alpha_5 x + \alpha_6 y \end{aligned}\right\} \tag{5.3}$$

Since these displacements are both linear in x and y, displacement continuity is ensured along the interface between adjoining elements for any nodal displacement. Displacement continuity is discussed more fully in Chapters 6 and 7.

Equation 5.3 can be written in matrix form as

$$\{\delta(x, y)\} = \begin{Bmatrix} u \\ v \end{Bmatrix} = \begin{bmatrix} 1 & x & y & 0 & 0 & 0 \\ 0 & 0 & 0 & 1 & x & y \end{bmatrix} \begin{Bmatrix} \alpha_1 \\ \alpha_2 \\ \alpha_3 \\ \alpha_4 \\ \alpha_5 \\ \alpha_6 \end{Bmatrix} \tag{5.4}$$

or briefly as

$$\{\delta(x, y)\} = [f(x, y)]\{\alpha\} \tag{II}$$

Step III: Express state of displacement $\{\delta(x, y)\}$ within element in terms of nodal displacements $\{\delta^e\}$

This step is achieved by substituting the values of the nodal coordinates into equation II (see also equation 5.4) and therefore obtaining expressions for the unknown coefficients. For example at node 1,

$$\{\delta_1\} = \delta(x_1, y_1)\} = [f(x_1, y_1)]\{\alpha\}$$

where $[f(x_1, y_1)]$ is given in equation 5.5a.

$$\{\delta_1\} = \begin{bmatrix} 1 & x_1 & y_1 & 0 & 0 & 0 \\ 0 & 0 & 0 & 1 & x_1 & y_1 \end{bmatrix} \{\alpha\} \qquad (5.5a)$$

Similarly for nodes 2 and 3 equations 5.5b and 5.5c are obtained.

$$\{\delta_2\} = \begin{bmatrix} 1 & x_2 & y_2 & 0 & 0 & 0 \\ 0 & 0 & 0 & 1 & x_2 & y_2 \end{bmatrix} \{\alpha\} \qquad (5.5b)$$

$$\{\delta_3\} = \begin{bmatrix} 1 & x_3 & y_3 & 0 & 0 & 0 \\ 0 & 0 & 0 & 1 & x_3 & y_3 \end{bmatrix} \{\alpha\} \qquad (5.5c)$$

Combining the displacement vectors for the element, equation 5.5d is obtained in general terms

$$\{\delta^e\} = \begin{Bmatrix} \{\delta_1\} \\ \{\delta_2\} \\ \{\delta_3\} \end{Bmatrix} = \begin{Bmatrix} \{\delta(x_1, y_1)\} \\ \{\delta(x_2, y_2)\} \\ \{\delta(x_3, y_3)\} \end{Bmatrix} = \begin{bmatrix} [f(x_1, y_1)] \\ [f(x_2, y_2)] \\ [f(x_3, y_3)] \end{bmatrix} \{\alpha\} \qquad (5.5d)$$

Substituting equations 5.5a, b and c into equation 5.5d, equation 5.6 is obtained.

$$\{\delta^e\} = \begin{Bmatrix} \{\delta_1\} \\ \{\delta_2\} \\ \{\delta_3\} \end{Bmatrix} = \begin{bmatrix} 1 & x_1 & y_1 & 0 & 0 & 0 \\ 0 & 0 & 0 & 1 & x_1 & y_1 \\ 1 & x_2 & y_2 & 0 & 0 & 0 \\ 0 & 0 & 0 & 1 & x_2 & y_2 \\ 1 & x_3 & y_3 & 0 & 0 & 0 \\ 0 & 0 & 0 & 1 & x_3 & y_3 \end{bmatrix} \begin{Bmatrix} \alpha_1 \\ \alpha_2 \\ \alpha_3 \\ \alpha_4 \\ \alpha_5 \\ \alpha_6 \end{Bmatrix} \qquad (5.6)$$

which may be written as

$$\{\delta^e\} = [A]\{\alpha\} \qquad (5.7)$$

where

$$[A] = \begin{bmatrix} 1 & x_1 & y_1 & 0 & 0 & 0 \\ 0 & 0 & 0 & 1 & x_1 & y_1 \\ 1 & x_2 & y_2 & 0 & 0 & 0 \\ 0 & 0 & 0 & 1 & x_2 & y_2 \\ 1 & x_3 & y_3 & 0 & 0 & 0 \\ 0 & 0 & 0 & 1 & x_3 & y_3 \end{bmatrix} \tag{5.8}$$

It should be noted that all the terms in matrix $[A]$ are known since they simply consist of the co-ordinates of the element nodes.

The unknown polynomial coefficients $\{\alpha\}$ are now determined from equation 5.7 by inverting the matrix of coefficients $[A]$ to yield

$$\{\alpha\} = [A]^{-1}\{\delta^e\} \tag{5.9}$$

To obtain equation 5.9 from equation 5.7 $[A]$ must be inverted. For the small matrix in this example, this can be done algebraically, but for larger matrices the inversion must be performed numerically by the computer.

In the present case, to obtain $[A]^{-1}$, $[A]$ can be rearranged to give two blocks of 3×3. On inverting a 3×3 block and again rearranging, $[A]^{-1}$ is found to be as shown in equation 5.10 (p. 62).

Using equations 5.8 and 5.10 the reader can verify that

$$[A]^{-1}[A] = [I]$$

Thus in equation 5.9 the desired relationship between the unknown coefficients $\{\alpha\}$ and the nodal displacements $\{\delta^e\}$ has been obtained.

By using equation II the displacements $\{\delta(x, y)\}$ at any point (x, y) within the element can now be determined in terms of the nodal displacements $\{\delta^e\}$. Equation II states that $\{\delta(x, y)\} = [f(x, y)] \{\alpha\}$. Hence substituting for $\{\alpha\}$ from equation 5.9,

$$\{\delta(x, y)\} = [f(x, y)][A]^{-1}\{\delta^e\} \tag{III}$$

$$[A]^{-1} = \frac{1}{2\Delta}
\begin{bmatrix}
x_2y_3 - x_3y_2 & 0 & -x_1y_3 + x_3y_1 & 0 & x_1y_2 - x_2y_1 & 0 \\
y_2 - y_3 & 0 & y_3 - y_1 & 0 & y_1 - y_2 & 0 \\
x_3 - x_2 & 0 & x_1 - x_3 & 0 & x_2 - x_1 & 0 \\
0 & x_2y_3 - x_3y_2 & 0 & -x_1y_3 + x_3y_1 & 0 & x_1y_2 - x_2y_1 \\
0 & y_2 - y_3 & 0 & y_3 - y_1 & 0 & y_1 - y_2 \\
0 & x_3 - x_2 & 0 & x_1 - x_3 & 0 & x_2 - x_1
\end{bmatrix}$$

$$(5.10)$$

where $2\Delta = \det \begin{vmatrix} 1 & x_1 & y_1 \\ 1 & x_2 & y_2 \\ 1 & x_3 & y_3 \end{vmatrix}$

$$= (x_2y_3 - x_3y_2) - (x_1y_3 - x_3y_1) + (x_1y_2 - x_2y_1)$$
$$= 2 \times \text{area of triangular element.}$$

Step IV: Relate strains $\{\varepsilon(x, y)\}$ at any point in element to displacements $\{\delta(x, y)\}$ and hence to nodal displacements $\{\delta^e\}$

For plane stress and plane strain problems the strain vector $\{\varepsilon(x, y)\}$ is as given in equation 5.11.

$$\{\varepsilon(x, y)\} = \begin{Bmatrix} \varepsilon_x \\ \varepsilon_y \\ \gamma_{xy} \end{Bmatrix} \qquad (5.11)$$

where ε_x and ε_y are the direct strains and γ_{xy} is the shearing strain.

From the theory of elasticity[1] the following relationship between strains ε and displacements u and v is known.

$$\left. \begin{aligned} \varepsilon_x &= \frac{\partial u}{\partial x} \\[2mm] \varepsilon_y &= \frac{\partial v}{\partial y} \\[2mm] \gamma_{xy} &= \frac{\partial u}{\partial y} + \frac{\partial v}{\partial x} \end{aligned} \right\} \qquad (5.12)$$

Substituting for u and v from equation 5.3,

$$\varepsilon_x = \frac{\partial}{\partial x}(\alpha_1 + \alpha_2 x + \alpha_3 y) \qquad\qquad = \alpha_2$$

$$\varepsilon_y = \frac{\partial}{\partial y}(\alpha_4 + \alpha_5 x + \alpha_6 y) \qquad\qquad = \alpha_6$$

$$\gamma_{xy} = \frac{\partial}{\partial y}(\alpha_1 + \alpha_2 x + \alpha_3 y) + \frac{\partial}{\partial x}(\alpha_4 + \alpha_5 x + \alpha_6 y) = \alpha_3 + \alpha_5$$

Thus

$$\{\varepsilon(x, y)\} = \begin{Bmatrix} \varepsilon_x \\ \varepsilon_y \\ \gamma_{xy} \end{Bmatrix} = \begin{Bmatrix} \alpha_2 \\ \alpha_6 \\ \alpha_3 + \alpha_5 \end{Bmatrix}$$

or

$$\{\varepsilon(x, y)\} = \left\{\begin{array}{c} \varepsilon_x \\ \varepsilon_y \\ \gamma_{xy} \end{array}\right\} = \begin{bmatrix} 0 & 1 & 0 & 0 & 0 & 0 \\ 0 & 0 & 0 & 0 & 0 & 1 \\ 0 & 0 & 1 & 0 & 1 & 0 \end{bmatrix} \left\{\begin{array}{c} \alpha_1 \\ \alpha_2 \\ \alpha_3 \\ \alpha_4 \\ \alpha_5 \\ \alpha_6 \end{array}\right\} \qquad (5.13)$$

or simply

$$\{\varepsilon(x, y)\} = [C]\{\alpha\} \qquad (5.14)$$

Substituting $[A]^{-1}\{\delta^e\}$ for $\{\alpha\}$ from equation 5.9,

$$\{\varepsilon(x, y)\} = [C][A]^{-1}\{\delta^e\} \qquad (5.15)$$

which may be written as

$$\{\varepsilon(x, y)\} = [B]\{\delta^e\} \qquad \text{(IV)}$$

where

$$[B] = [C][A]^{-1} \qquad (5.16)$$

Thus equation IV relates the strains $\{\varepsilon(x, y)\}$ at any point within the element to the nodal displacements $\{\delta^e\}$.

The matrix $[B]$ can be obtained explicitly by multiplying matrix $[C]$ from equation 5.13 with $[A]^{-1}$ as obtained in equation 5.10. Performing the algebra

$$[B] = \frac{1}{2\Delta} \begin{bmatrix} y_2 - y_3 & 0 & y_3 - y_1 & 0 & y_1 - y_2 & 0 \\ 0 & x_3 - x_2 & 0 & x_1 - x_3 & 0 & x_2 - x_1 \\ x_3 - x_2 & y_2 - y_3 & x_1 - x_3 & y_3 - y_1 & x_2 - x_1 & y_1 - y_2 \end{bmatrix}$$

$$(5.17)$$

where 2Δ is as defined after equation 5.10.

Step V : Relate internal stresses $\{\sigma(x, y)\}$ to strains $\{\varepsilon(x, y)\}$ and to nodal displacements $\{\delta^e\}$

For a plane elasticity problem, the state of stress $\{\sigma(x, y)\}$ at any point may be represented by three stress components σ_x, σ_y and τ_{xy}.

$$\{\sigma(x, y)\} = \begin{Bmatrix} \sigma_x \\ \sigma_y \\ \tau_{xy} \end{Bmatrix} \tag{5.18}$$

where σ_x and σ_y are direct stresses and τ_{xy} is the shearing stress. The corresponding strain components are given in equation 5.11. These stress and strain components are related by the $[D]$ matrix, where $\{\sigma(x, y)\} = [D]\{\varepsilon(x, y)\}$. For plane elasticity problems $[D]$ is a 3×3 matrix, the terms of which depend upon whether the problem is one of plane stress or plane strain.

For a plane stress problem (i.e. with zero stress σ_z normal to the plane) the following relationships exist between the strains and stresses.[1]

$$\varepsilon_x = \frac{\sigma_x}{E} - \frac{\nu \sigma_y}{E}$$

$$\varepsilon_y = \frac{-\nu \sigma_x}{E} + \frac{\sigma_y}{E}$$

$$\gamma_{xy} = \frac{\tau_{xy}}{G} = \frac{2(1+\nu)}{E} \tau_{xy}$$

where E is Young's modulus of elasticity, G is the modulus of rigidity and ν is Poisson's ratio.

$$\{\varepsilon(x, y)\} = \begin{Bmatrix} \varepsilon_x \\ \varepsilon_y \\ \gamma_{xy} \end{Bmatrix} = \frac{1}{E} \begin{bmatrix} 1 & -\nu & 0 \\ -\nu & 1 & 0 \\ 0 & 0 & 2(1+\nu) \end{bmatrix} \begin{Bmatrix} \sigma_x \\ \sigma_y \\ \tau_{xy} \end{Bmatrix} \tag{5.19}$$

By re-arranging equation 5.19 it can be shown that

$$\{\sigma(x, y)\} = \begin{Bmatrix} \sigma_x \\ \sigma_y \\ \tau_{xy} \end{Bmatrix} = \frac{E}{1-v^2} \begin{bmatrix} 1 & v & 0 \\ v & 1 & 0 \\ 0 & 0 & \dfrac{1-v}{2} \end{bmatrix} \begin{Bmatrix} \varepsilon_x \\ \varepsilon_y \\ \gamma_{xy} \end{Bmatrix} \tag{5.20}$$

or

$$\{\sigma(x, y)\} = [D]\{\varepsilon(x, y)\} \tag{5.21}$$

Substituting for $\{\varepsilon(x, y)\}$ from equation IV, the following relationship between the element stresses and the nodal displacements is obtained.

Plane stress: $\qquad \{\sigma(x, y)\} = [D][B]\{\delta^e\} \tag{Va}$

For plane strain (with zero strain normal to the plane) the following relationships exist in elastic theory[1] between strains and stresses.

$$\begin{aligned}
\varepsilon_x &= \frac{\sigma_x}{E} - \frac{v\sigma_y}{E} - \frac{v\sigma_z}{E} \\
\varepsilon_y &= \frac{-v\sigma_x}{E} + \frac{\sigma_y}{E} - \frac{v\sigma_z}{E} \\
\varepsilon_z &= \frac{-v\sigma_x}{E} - \frac{v\sigma_y}{E} + \frac{\sigma_z}{E} = 0 \text{ (plane strain)} \\
\gamma_{xy} &= \frac{\tau_{xy}}{G} = \frac{2(1+v)}{E}\tau_{xy}
\end{aligned} \right\} \tag{5.22}$$

On eliminating σ_z and solving for σ_x, σ_y and τ_{xy},

$$\begin{Bmatrix} \sigma_x \\ \sigma_y \\ \tau_{xy} \end{Bmatrix} = \frac{E(1-v)}{(1+v)(1-2v)} \begin{bmatrix} 1 & \dfrac{v}{1-v} & 0 \\ \dfrac{v}{1-v} & 1 & 0 \\ 0 & 0 & \dfrac{1-2v}{2(1-v)} \end{bmatrix} \begin{Bmatrix} \varepsilon_x \\ \varepsilon_y \\ \gamma_{xy} \end{Bmatrix} \tag{5.23}$$

or

$$\{\sigma(x, y)\} = [D]\{\varepsilon(x, y)\} \tag{5.24}$$

Again substituting for $\{\varepsilon(x, y)\}$ from equation IV,

Plane strain: $$\{\sigma(x, y)\} = [D][B]\{\delta^e\} \tag{Vb}$$

Clearly the $[D]$ matrix in equation V is different for the cases of plane stress and plane strain. However, for convenience for plane elasticity problems the $[D]$ matrix can be expressed as

$$[D] = \begin{bmatrix} d_{11} & d_{12} & 0 \\ d_{21} & d_{22} & 0 \\ 0 & 0 & d_{33} \end{bmatrix}$$

where, for plane stress (equation 5.20)

$$d_{11} = d_{22} = E/(1-v^2),$$
$$d_{12} = d_{21} = vE/(1-v^2),$$
$$d_{33} = E/2(1+v);$$

and for plane strain (equation 5.23)

$$d_{11} = d_{22} = E(1-v)/(1+v)(1-2v),$$
$$d_{12} = d_{21} = vE/(1+v)(1-2v),$$
$$d_{33} = E/2(1+v)$$

Step VI: Replace internal stresses $\{\sigma(x, y)\}$ with statically equivalent nodal forces $\{F^e\}$, relate nodal forces to nodal displacements $\{\delta^e\}$ and hence obtain element stiffness matrix $[K^e]$

The details in this step are exactly the same as those which were developed in Chapter 3 and the final result is as before,

$$\{F^e\} = [\int [B]^T [D][B] \, d(\text{vol})]\{\delta^e\} \tag{VI}$$

For plane elasticity, $[B]$ has been obtained explicitly for the triangle and is given in equation 5.17. The $[D]$ matrix depends upon whether the problem being considered is one of plane stress (equation 5.20) or plane strain (equation 5.23).

It should be noted that, since the matrices $[B]$ and $[D]$ contain only constant terms, they can be taken outside the integration in equation VI, leaving only $\int d(\text{vol})$ which, in the case of an element of constant thickness equals the area of the triangle Δ multiplied by its thickness t. Thus for an element of constant thickness,

$$\{F^e\} = [[B]^T[D][B]\,\Delta t]\{\delta^e\} \qquad (5.25)$$

and

$$2\Delta = \det \begin{vmatrix} 1 & x_1 & y_1 \\ 1 & x_2 & y_2 \\ 1 & x_3 & y_3 \end{vmatrix} \qquad (5.26)$$

The element stiffness matrix is therefore

$$[K^e] = [[B]^T[D][B]\,\Delta t] \qquad (5.27)$$

The terms of $[K^e]$ can be obtained explicitly from equation 5.27 by multiplying out the matrices $[B]^T$, $[D]$ and $[B]$, and are given below for the triangular plane elasticity element.

First the product $[D][B]$ is obtained

$$[D][B] = \frac{1}{2\Delta}\begin{bmatrix} d_{11}(y_2-y_3) & d_{12}(x_3-x_2) & d_{11}(y_3-y_1) & d_{12}(x_1-x_3) & d_{11}(y_1-y_2) & d_{12}(x_2-x_1) \\ d_{21}(y_2-y_3) & d_{22}(x_3-x_2) & d_{21}(y_3-y_1) & d_{22}(x_1-x_3) & d_{21}(y_1-y_2) & d_{22}(x_2-x_1) \\ d_{33}(x_3-x_2) & d_{33}(y_2-y_3) & d_{33}(x_1-x_3) & d_{33}(y_3-y_1) & d_{33}(x_2-x_1) & d_{33}(y_1-y_2) \end{bmatrix}$$

$$(5.28)$$

where Δ, the area of the triangle, is as given in equation 5.26. To obtain the element stiffness matrix $[K^e]$ as given in equation 5.27 this product must be pre-multiplied by the transpose of matrix $[B]$ from equation 5.17 and also by Δt, the volume of the element.

$$[B]^T = \frac{1}{2\Delta}\begin{bmatrix} y_2-y_3 & 0 & x_3-x_2 \\ 0 & x_3-x_2 & y_2-y_3 \\ y_3-y_1 & 0 & x_1-x_3 \\ 0 & x_1-x_3 & y_3-y_1 \\ y_1-y_2 & 0 & x_2-x_1 \\ 0 & x_2-x_1 & y_1-y_2 \end{bmatrix} \qquad (5.29)$$

Thus since $[K^e]=[[B]^T[D][B]\,\Delta t]$, equation 5.30, shown on p. 69, is obtained.

$$
[K^e] = \frac{t}{4\Delta}
$$

$$
\begin{bmatrix}
d_{11}(y_2-y_3)^2 + d_{33}(x_3-x_2)^2 & d_{12}(x_3-x_2)(y_2-y_3) + d_{33}(x_3-x_2)(y_2-y_3) & d_{11}(y_2-y_3)(y_3-y_1) + d_{33}(x_3-x_2)(x_1-x_3) & d_{12}(x_1-x_3)(y_2-y_3) + d_{33}(x_3-x_2)(y_3-y_1) & d_{11}(y_1-y_2)(y_2-y_3) + d_{33}(x_2-x_1)(x_3-x_2) & d_{12}(x_2-x_1)(y_2-y_3) + d_{33}(x_3-x_2)(y_1-y_2) \\[4pt]
d_{21}(x_3-x_2)(y_2-y_3) + d_{33}(x_3-x_2)(y_2-y_3) & d_{22}(x_3-x_2)^2 + d_{33}(y_2-y_3)^2 & d_{12}(y_3-y_1)(x_3-x_2) + d_{33}(x_1-x_3)(y_2-y_3) & d_{22}(x_3-x_2)(x_1-x_3) + d_{33}(y_2-y_3)(y_3-y_1) & d_{21}(x_3-x_2)(y_1-y_2) + d_{33}(x_2-x_1)(y_2-y_3) & d_{22}(x_2-x_1)(x_3-x_2) + d_{33}(y_1-y_2)(y_2-y_3) \\[4pt]
d_{11}(y_2-y_3)(y_3-y_1) + d_{33}(x_1-x_3)(x_3-x_2) & d_{12}(x_3-x_2)(y_3-y_1) + d_{33}(x_1-x_3)(y_2-y_3) & d_{11}(y_3-y_1)^2 + d_{33}(x_1-x_3)^2 & d_{12}(x_1-x_3)(y_3-y_1) + d_{33}(x_1-x_3)(y_3-y_1) & d_{11}(y_1-y_2)(y_3-y_1) + d_{33}(x_1-x_3)(x_2-x_1) & d_{12}(x_2-x_1)(y_3-y_1) + d_{33}(x_1-x_3)(y_1-y_2) \\[4pt]
d_{21}(x_1-x_3)(y_2-y_3) + d_{33}(x_3-x_2)(y_3-y_1) & d_{22}(x_1-x_3)(x_3-x_2) + d_{33}(y_2-y_3)(y_3-y_1) & d_{12}(x_1-x_3)(y_3-y_1) + d_{33}(x_1-x_3)(y_3-y_1) & d_{22}(x_1-x_3)^2 + d_{33}(y_3-y_1)^2 & d_{21}(x_1-x_3)(y_1-y_2) + d_{33}(x_2-x_1)(y_3-y_1) & d_{22}(x_1-x_3)(x_2-x_1) + d_{33}(y_3-y_1)(y_1-y_2) \\[4pt]
d_{11}(y_1-y_2)(y_2-y_3) + d_{33}(x_2-x_1)(x_3-x_2) & d_{12}(x_3-x_2)(y_1-y_2) + d_{33}(x_2-x_1)(y_2-y_3) & d_{11}(y_1-y_2)(y_3-y_1) + d_{33}(x_1-x_3)(x_2-x_1) & d_{12}(x_1-x_3)(y_1-y_2) + d_{33}(x_2-x_1)(y_3-y_1) & d_{11}(y_1-y_2)^2 + d_{33}(x_2-x_1)^2 & d_{12}(x_2-x_1)(y_1-y_2) + d_{33}(x_2-x_1)(y_1-y_2) \\[4pt]
d_{21}(x_2-x_1)(y_2-y_3) + d_{33}(x_3-x_2)(y_1-y_2) & d_{22}(x_2-x_1)(x_3-x_2) + d_{33}(y_1-y_2)(y_2-y_3) & d_{12}(x_2-x_1)(y_3-y_1) + d_{33}(x_1-x_3)(y_1-y_2) & d_{22}(x_2-x_1)(x_1-x_3) + d_{33}(y_3-y_1)(y_1-y_2) & d_{21}(x_2-x_1)(y_1-y_2) + d_{33}(x_2-x_1)(y_1-y_2) & d_{22}(x_2-x_1)^2 + d_{33}(y_1-y_2)^2
\end{bmatrix}
$$

$$(5.30)$$

The explicit form for $[K^e]$ is obtained in equation 5.30 from which it can be noted that this matrix is symmetrical. The cases of plane stress and plane strain can be obtained by substituting for d_{ij} from equations 5.20 and 5.23. It is simpler in practice to perform the matrix multiplications of equation 5.27 numerically in the computer.

Having obtained the element stiffness matrix the nodal displacements may now be calculated from the nodal forces.

Step VII: Establish stress–displacement matrix $[H]$

The final step is to determine the element stresses from the element nodal displacements. The relationship in equation V enables this to be done.

$$\{\sigma(x, y)\} = [D][B]\{\delta^e\} \tag{V}$$

or

$$\{\sigma(x, y)\} = [H]\{\delta^e\} \tag{VII}$$

For the case of plane elasticity, the product $[H] = [D][B]$ is given explicitly in equation 5.28.

In the cases of the plane stress and plane strain triangular element, it is seen that the strain is constant throughout the element—this element is thus usually referred to as the constant strain triangle. For convenience it is usual to plot the stresses at the centroid of the element.

Thus the same seven basic steps described in Chapter 3 for the beam element have again enabled the nodal displacements and hence the element stresses to be obtained.

Applications

Example of use of triangular element

Consider the deep beam of rectangular section shown in Figure 5.5 which is simply supported at its ends and carries the uniformly distributed loading shown. This simple example has been chosen because an exact solution has been obtained using the theory of

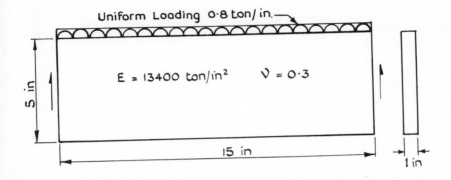

Fig. 5.5. Simply supported deep beam under uniform loading

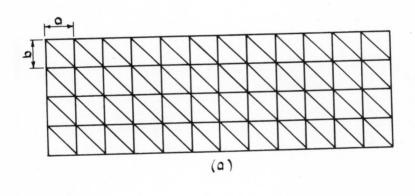

(a)

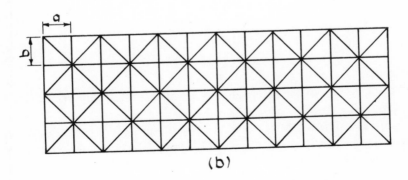

(b)

Fig. 5.6. Triangular element idealisations

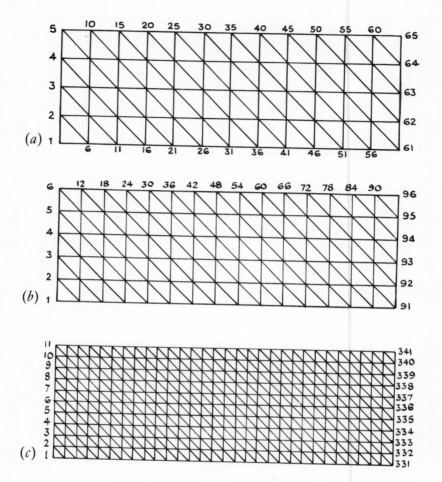

Fig. 5.7. Idealisations used for deep beam. *a* 65 nodes, 96 elements. *b* 96 nodes, 150 elements. *c* 341 nodes, 600 elements

elasticity.[1] It is thus possible to compare the finite element solution with the exact solution when different finite element idealisations are employed.

Figure 5.6 shows two possible idealisations that could be employed using triangular elements. Since the shape of the structure is rectangular a regular mesh can be used throughout. It is important at all times to keep the aspect ratio $a:b$ (see Figure 5.6*a*) as near to unity as possible, i.e. to avoid long thin triangles since their use reduces

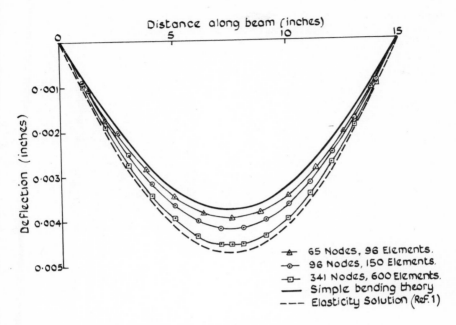

Fig. 5.8. Deflection of simply supported deep beam with uniform loading

accuracy. It has been reported[2] that the type of mesh shown in Figure 5.6a is to be preferred since it results in smaller errors. It also has the advantage that automatic generation of the mesh can be accomplished easily.

Figure 5.7 shows three idealisations for which solutions have been obtained. In Figure 5.7a there are four divisions through the depth, while in Figures 5.7b and 5.7c there are five and ten divisions through the depth. The simply-supported condition is enforced by specifying zero vertical displacement of all the nodes at the extreme ends of the beam. Figure 5.8 shows the deflected shape of the beam using simple bending theory which ignores shear deformations (shown by a broken line), the theory of elasticity solution[1] (shown by a continuous continuous line), the theory of elasticity solution[1] (shown by a broken employed, plotting the centre-line values in each case. Figure 5.9 shows the corresponding information for the variation in longitudinal and transverse stresses. The stresses obtained by the finite element method are constant within each triangular element, and the values are plotted at the centroids of the elements.

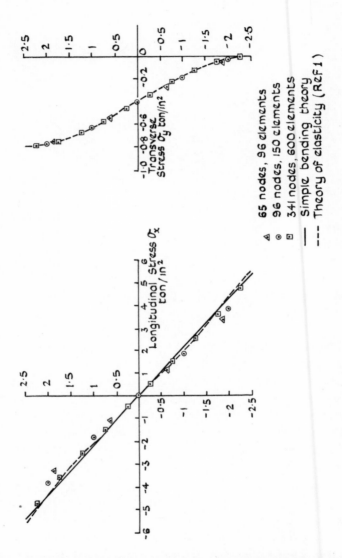

Fig. 5.9. Distribution of longitudinal and transverse stress in simply supported
deep beam with uniform loading

In each case it is seen that the finite element results under-estimate the exact answers but that, as the finite element subdivision is refined, so the answers approach the exact solutions. As stated in Chapter 4, it is always recommended that any problem should be solved using several idealisations to check that the answers converge as the number of elements is increased.

Conclusions

The element developed in this chapter is often referred to as the constant strain element since, as shown in step 4, the strain components are constant throughout this element. This is a direct consequence of the chosen displacement functions (equation 5.3) which are linear polynomials and, as given in the theory of elasticity[1] by equation 5.12, the strains are the derivatives of the displacements. Thus there are usually discontinuities of some or all the strain components (ε_x, ε_y and γ_{xy}) at element boundaries. However, the displacement functions (equation 5.3) do ensure continuity of displacement between elements at their boundaries. No gaps can occur between the edges of adjacent deformed elements because the displacements permitted are expressed linearly in terms of nodal displacements (step III), so that for adjacent elements having two common nodes, the edges joining these nodes remain linear in both elements.

It may be easier to visualise this by constructing a three-dimensional model of the displacement functions of equation 5.3. Measuring u and v vertically from the x–y plane, each displacement function will trace a surface above the x–y plane. This surface will be planar, giving an inclined triangular plane for each element because u and v are linear polynomials in x and y. In adjacent elements the planar surfaces will in general be inclined at different angles and are such that they always meet along the edges of neighbouring elements. Hence for both u and v the surfaces are continuous, being made up of tilted planar triangles.

The direct strain in an element in any particular direction is the slope of the displacement surface in that direction, and is thus constant within the element. At the boundary there is generally a discontinuity in the slope. In view of this fact it is necessary to use a fine mesh of these elements in regions having high stress gradients. In plotting results the stresses and strains are frequently plotted at the element centroids, as in Figure 5.9, though other techniques, such as calculating nodal values by averaging the stresses or strains in elements meeting at the node with the possibility of using weighting factors in some cases, have been used.

The triangular element has the advantages of simplicity in use and the ability to fit irregular boundaries. It has been used successfully in a wide range of practical problems.

References

1. TIMOSHENKO, S. P. and GOODIER, J. N. *Theory of elasticity*. New York, McGraw-Hill Book Co. 1951.
2. WALZ, J. E., FULTON, R. E. and CYRUS, N. J. Accuracy and convergence of finite element approximations. *Proceedings of the Second Conference on Matrix Methods in Structural Mechanics at Wright-Patterson Air Force Base, Ohio, 1968.*

CHAPTER SIX

Rectangular Finite Element for Plane Elasticity

In this chapter the seven basic steps are employed to derive the stiffness characteristics for a rectangular element in plane stress and plane strain. This element can be applied to the solution of problems similar to those that can be solved using the triangular element discussed in Chapter 5.

Derivation of rectangular element stiffness matrix

Step I: Choose suitable co-ordinates and number nodes

Let the rectangular element have sides of lengths a and b, and a thickness t as shown in Figure 6.1a. The node numbering system shown in Figure 6.1a is employed. Other systems could have been chosen but the one used here has been found to be convenient.

Since for a plane elasticity problem, the element has two degrees of freedom at each node, each element has eight degrees of freedom. Figure 6.1b shows the eight unknown displacements and Figure 6.1c the corresponding nodal forces. Using matrix notation, the displacements at node 1 may then be written as

$$\{\delta_1\} = \begin{Bmatrix} u_1 \\ v_1 \end{Bmatrix}$$

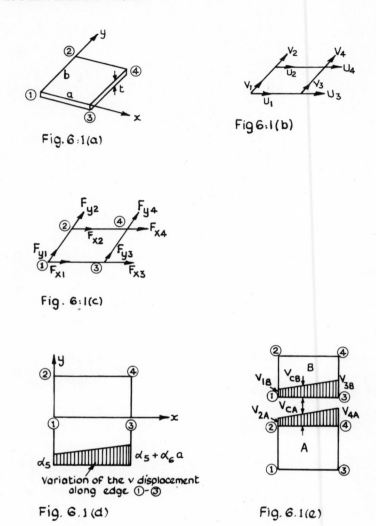

Fig. 6:1(a)

Fig 6:1(b)

Fig. 6:1(c)

Variation of the v displacement
along edge ①-③

Fig. 6.1(d)

Fig. 6.1(e)

Fig. 6.1. *a, b* Nodal displacements. *c* Nodal forces. *d, e*

and the forces at node 1 may be written as

$$\{F_1\} = \begin{Bmatrix} F_{x1} \\ F_{y1} \end{Bmatrix}$$

The complete displacement and force vectors for the element are then as shown in equations 6.1 and 6.2.

$$\{\delta^e\} = \begin{Bmatrix} \{\delta_1\} \\ \{\delta_2\} \\ \{\delta_3\} \\ \{\delta_4\} \end{Bmatrix} = \begin{Bmatrix} u_1 \\ v_1 \\ \hline u_2 \\ v_2 \\ \hline u_3 \\ v_3 \\ \hline u_4 \\ v_4 \end{Bmatrix} \tag{6.1}$$

$$\{F^e\} = \begin{Bmatrix} \{F_1\} \\ \{F_2\} \\ \{F_3\} \\ \{F_4\} \end{Bmatrix} = \begin{Bmatrix} F_{x1} \\ F_{y1} \\ \hline F_{x2} \\ F_{y2} \\ \hline F_{x3} \\ F_{y3} \\ \hline F_{x4} \\ F_{y4} \end{Bmatrix} \tag{6.2}$$

Each of these vectors contains eight terms so that the element stiffness matrix $[K^e]$ is now an 8×8 matrix.

$$\{F^e\} = [K^e]\{\delta^e\} \tag{I}$$

Step II: Choose displacement function $[f(x, y)]$ that defines displacement $\{\delta(x, y)\}$ at any point in element

In a plane elasticity problem the state of displacement at any point (x, y) within the element may, as in Chapter 5, be represented by two components, i.e.

$$\{\delta(x, y)\} = \begin{Bmatrix} u \\ v \end{Bmatrix}$$

Since the element has eight degrees of freedom, eight unknown coefficients must be involved in the polynomial representing the displacement pattern. Two such suitable functions are given in equations 6.3.

$$u = \alpha_1 + \alpha_2 x + \alpha_3 y + \alpha_4 xy \Big\}$$
$$v = \alpha_5 + \alpha_6 x + \alpha_7 y + \alpha_8 xy \Big\}$$

(6.3)

It may be noted that when x is a constant both u and v vary linearly with y, and similarly when y is a constant both displacements vary linearly with x. The displacements thus vary linearly along each side of the rectangular element.

For example, along edge ①–③ $y=0$, so that $u=\alpha_1 + \alpha_2 x$ and $v=\alpha_5 + \alpha_6 x$. The v displacement thus varies linearly along this edge from a value of α_5 at node ① (where $x=0$) to a value of $\alpha_5 + \alpha_6 a$ at node ③ (where $x=a$) (see Figure 6.1d). Since identical displacements are imposed at the nodes, i.e. compatibility of nodal displacements is ensured, the same displacements exist in adjacent elements at all points along the interface. This may be illustrated by considering two adjacent elements A and B (Figure 6.1e). During the solution it is specified that

$$v_{2A} = v_{1B} \quad \text{and} \quad v_{4A} = v_{3B}$$

Consequently,

$$v_{cA} = v_{cB}$$

The choice of this function thus ensures full displacement continuity in the solution.

Writing equation 6.3 in matrix form

$$\begin{Bmatrix} u \\ v \end{Bmatrix} = \begin{bmatrix} 1 & x & y & xy & 0 & 0 & 0 & 0 \\ 0 & 0 & 0 & 0 & 1 & x & y & xy \end{bmatrix} \begin{Bmatrix} \alpha_1 \\ \alpha_2 \\ \alpha_3 \\ \alpha_4 \\ \alpha_5 \\ \alpha_6 \\ \alpha_7 \\ \alpha_8 \end{Bmatrix}$$

(6.4)

which may be summarised as in general equation II as

$$\{\delta(x, y)\} = [f(x, y)]\{\alpha\} \tag{II}$$

Step III: Express state of displacement $\{\delta(x, y)\}$ at any point within element in terms of nodal displacements $\{\delta^e\}$

As previously carried out in Chapter 5, this step is accomplished by substituting the values of the nodal co-ordinates into equation II and then solving for $\{\alpha\}$. Substituting the co-ordinates of the nodes into $[f(x, y)]$, at node 1, $x_1 = 0$, $y_1 = 0$, therefore,

$$[f(x_1, y_1)] = \begin{bmatrix} 1 & 0 & 0 & 0 & 0 & 0 & 0 & 0 \\ 0 & 0 & 0 & 0 & 1 & 0 & 0 & 0 \end{bmatrix}$$

at node 2, $x_2 = 0$, $y_2 = b$, therefore,

$$[f(x_2, y_2)] = \begin{bmatrix} 1 & 0 & b & 0 & 0 & 0 & 0 & 0 \\ 0 & 0 & 0 & 0 & 1 & 0 & b & 0 \end{bmatrix}$$

at node 3, $x_3 = a$, $y_3 = 0$, therefore,

$$[f(x_3, y_3)] = \begin{bmatrix} 1 & a & 0 & 0 & 0 & 0 & 0 & 0 \\ 0 & 0 & 0 & 0 & 1 & a & 0 & 0 \end{bmatrix}$$

at node 4, $x_4 = a$, $y_4 = b$, therefore,

$$[f(x_4, y_4)] = \begin{bmatrix} 1 & a & b & ab & 0 & 0 & 0 & 0 \\ 0 & 0 & 0 & 0 & 1 & a & b & ab \end{bmatrix}$$

Thus using equation 6.4,

$$\{\delta^e\} = \begin{Bmatrix} \{\delta_1\} \\ \{\delta_2\} \\ \{\delta_3\} \\ \{\delta_4\} \end{Bmatrix} = \begin{Bmatrix} \{\delta(x_1, y_1)\} \\ \{\delta(x_2, y_2)\} \\ \{\delta(x_3, y_3)\} \\ \{\delta(x_4, y_4)\} \end{Bmatrix} = \begin{bmatrix} [f(x_1, y_1)] \\ [f(x_2, y_2)] \\ [f(x_3, y_3)] \\ [f(x_4, y_4)] \end{bmatrix} \{\alpha\} \tag{6.5}$$

On substituting for $[f(x_1, y_1)]$ etc, equation 6.5 results in equation 6.6.

$$\{\delta^e\} = \left[\begin{array}{cccccccc} 1 & 0 & 0 & 0 & 0 & 0 & 0 & 0 \\ 0 & 0 & 0 & 0 & 1 & 0 & 0 & 0 \\ \hline 1 & 0 & b & 0 & 0 & 0 & 0 & 0 \\ 0 & 0 & 0 & 0 & 1 & 0 & b & 0 \\ \hline 1 & a & 0 & 0 & 0 & 0 & 0 & 0 \\ 0 & 0 & 0 & 0 & 1 & a & 0 & 0 \\ \hline 1 & a & b & ab & 0 & 0 & 0 & 0 \\ 0 & 0 & 0 & 0 & 1 & a & b & ab \end{array}\right] \{\alpha\} \qquad (6.6)$$

and this may be summarised as

$$\{\delta^e\} = [A]\{\alpha\} \qquad (6.7)$$

where $\{\alpha\}$ is the vector of the unknown coefficients of the polynomial expression given in equation 6.3. Thus equation 6.7 defines the $[A]$ matrix for the particular case of a rectangular element in plane elasticity problems.

The unknown coefficients $\{\alpha\}$ are then obtained by pre-multiplying both sides of equation 6.7 by the inverse of $[A]$ yielding equation 6.8.

$$\{\alpha\} = [A]^{-1}\{\delta^e\} \qquad (6.8)$$

As in Chapter 5, the inverse of the $[A]$ matrix can be obtained either numerically in the computer or algebraically. It should be remembered that this process of inversion is simply a convenient way of solving eight simultaneous equations so that the values of the eight unknown α coefficients can be expressed in terms of the u and v displacements of the nodes. The simultaneous equations represented in equation 6.6 are

$$u_1 = \alpha_1$$
$$v_1 = \alpha_5$$
$$u_2 = \alpha_1 + b\alpha_3$$
$$v_2 = \alpha_5 + b\alpha_7$$
$$u_3 = \alpha_1 + a\alpha_2$$
$$v_3 = \alpha_5 + a\alpha_6$$
$$u_4 = \alpha_1 + a\alpha_2 + b\alpha_3 + ab\alpha_4$$
$$v_4 = \alpha_5 + a\alpha_6 + b\alpha_7 + ab\alpha_8$$

The algebraic solution of these equations is very simple, having obtained the solution, equation 6.8 may be written in full as

$$
\begin{Bmatrix} \alpha_1 \\ \alpha_2 \\ \alpha_3 \\ \alpha_4 \\ \alpha_5 \\ \alpha_6 \\ \alpha_7 \\ \alpha_8 \end{Bmatrix}
=
\begin{bmatrix}
1 & 0 & 0 & 0 & 0 & 0 & 0 & 0 \\
-1/a & 0 & 0 & 0 & 1/a & 0 & 0 & 0 \\
-1/b & 0 & 1/b & 0 & 0 & 0 & 0 & 0 \\
1/ab & 0 & -1/ab & 0 & -1/ab & 0 & 1/ab & 0 \\
0 & 1 & 0 & 0 & 0 & 0 & 0 & 0 \\
0 & -1/a & 0 & 0 & 0 & 1/a & 0 & 0 \\
0 & -1/b & 0 & 1/b & 0 & 0 & 0 & 0 \\
0 & 1/ab & 0 & -1/ab & 0 & -1/ab & 0 & 1/ab
\end{bmatrix}
\begin{Bmatrix} u_1 \\ v_1 \\ u_2 \\ v_2 \\ u_3 \\ v_3 \\ u_4 \\ v_4 \end{Bmatrix}
$$

thus defining the inverse of the $[A]$ matrix, i.e. $[A]^{-1}$. The inverse obtained may be checked by carrying out the multiplication $[A]^{-1}$. $[A]$ to ensure that the product is equal to the identity matrix $[I]$.

Using equation 6.8 general equation II can be re-written to express $\{\delta(x, y)\}$ in terms of $\{\delta^e\}$ as

$$
\{\delta(x, y)\} = [f(x, y)][A]^{-1}\{\delta^e\} \tag{III}
$$

Step IV: Relate strains $\{\varepsilon(x, y)\}$ at any point to displacements $\{\delta(x, y)\}$ and hence to nodal displacements $\{\delta^e\}$

Now the relationship between strains and displacements at a point in a plane elasticity solution is obviously independent of the shape of element chosen. Hence equations 5.11 and 5.12 derived for the triangular element are also valid for the rectangular element now being considered. Thus

$$
\{\varepsilon(x, y)\} = \begin{Bmatrix} \varepsilon_x \\ \varepsilon_y \\ \gamma_{xy} \end{Bmatrix} \tag{5.11}
$$

and

$$\left.\begin{array}{l} \varepsilon_x = \partial u/\partial x \\ \varepsilon_y = \partial v/\partial y \\ \gamma_{xy} = \partial u/\partial y + \partial v/\partial x \end{array}\right\} \qquad (5.12)$$

Substituting for u and v from equation 6.3, the following expressions are obtained for the strains at any point within the element.

$$\varepsilon_x = \frac{\partial}{\partial x}(\alpha_1 + \alpha_2 x + \alpha_3 y + \alpha_4 xy) = \alpha_2 + \alpha_4 y$$

$$\varepsilon_y = \frac{\partial}{\partial y}(\alpha_5 + \alpha_6 x + \alpha_7 y + \alpha_8 xy) = \alpha_7 + \alpha_8 x$$

$$\gamma_{xy} = \frac{\partial}{\partial y}(\alpha_1 + \alpha_2 x + \alpha_3 y + \alpha_4 xy) + \frac{\partial}{\partial x}(\alpha_5 + \alpha_6 x + \alpha_7 y + \alpha_8 xy)$$

$$= \alpha_3 + \alpha_4 x + \alpha_6 + \alpha_8 y.$$

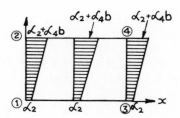

Fig. 6.2.

It is of particular interest to note that these strains vary linearly over the element. Considering the variation of ε_x as an example, this is seen to be independent of x and linearly dependent on y and thus varies over the element in the manner shown in Figure 6.2. Similarly ε_y is independent of y and linearly dependent on x whereas γ_{xy} varies linearly with both x and y.

Substituting these expressions for the strains into equation 5.11,

$$\{\varepsilon(x, y)\} = \begin{Bmatrix} \varepsilon_x \\ \varepsilon_y \\ \gamma_{xy} \end{Bmatrix} = \begin{Bmatrix} \alpha_2 + \alpha_4 y \\ \alpha_7 + \alpha_8 x \\ \alpha_3 + \alpha_4 x + \alpha_6 + \alpha_8 y \end{Bmatrix}$$

$$= \begin{bmatrix} 0 & 1 & 0 & y & 0 & 0 & 0 & 0 \\ 0 & 0 & 0 & 0 & 0 & 0 & 1 & x \\ 0 & 0 & 1 & x & 0 & 1 & 0 & y \end{bmatrix} \begin{Bmatrix} \alpha_1 \\ \alpha_2 \\ \alpha_3 \\ \alpha_4 \\ \alpha_5 \\ \alpha_6 \\ \alpha_7 \\ \alpha_8 \end{Bmatrix} \qquad (6.9)$$

Therefore

$$\{\varepsilon(x, y)\} = [C]\{\alpha\} \qquad (6.10)$$

On substituting for $\{\alpha\}$, using equation 6.8,

$$\{\varepsilon(x, y)\} = [C][A]^{-1}\{\delta^e\} \qquad (6.11)$$

or

$$\{\varepsilon(x, y)\} = [B]\{\delta^e\} \qquad \text{(IV)}$$

where

$$[B] = [C][A]^{-1} \qquad (6.12)$$

As shown previously in Chapter 5 equation IV relates the strains at any point to the nodal displacements as required.

Matrix $[B]$ is established from equation 6.12 exactly as in Chapter 5. Taking the $[C]$ matrix defined in equation 6.9 and the $[A]^{-1}$ matrix defined in equation 6.8, the $[B]$ matrix is obtained as

$$\begin{bmatrix} -\dfrac{1}{a}+\dfrac{y}{ab} & 0 & \dfrac{-y}{ab} & 0 & \dfrac{1}{a}-\dfrac{y}{ab} & 0 & \dfrac{y}{ab} & 0 \\[2ex] 0 & -\dfrac{1}{b}+\dfrac{x}{ab} & 0 & \dfrac{1}{b}-\dfrac{x}{ab} & 0 & -\dfrac{x}{ab} & 0 & \dfrac{x}{ab} \\[2ex] -\dfrac{1}{b}+\dfrac{x}{ab} & -\dfrac{1}{a}+\dfrac{y}{ab} & \dfrac{1}{b}-\dfrac{x}{ab} & -\dfrac{y}{ab} & -\dfrac{x}{ab} & \dfrac{1}{a}-\dfrac{y}{ab} & \dfrac{x}{ab} & \dfrac{y}{ab} \end{bmatrix}$$

Step V: Relate internal stresses $\{\sigma(x, y)\}$ to strains $\{\varepsilon(x, y)\}$ and to nodal displacements $\{\delta^e\}$

The relationship between stresses and strains given in equation 5.21 is clearly independent of the shape of element taken, so equations 5.18 to 5.24 are also valid for the rectangular element. The matrix $[D]$ required in general equation V, that is

$$\{\sigma(x, y)\} = [D][B]\{\delta^e\} \tag{V}$$

which gives the required relationship between stresses at any point and nodal displacements, is thus as defined in Chapter 5, namely

$$[D] = \begin{bmatrix} d_{11} & d_{12} & 0 \\ d_{21} & d_{22} & 0 \\ 0 & 0 & d_{33} \end{bmatrix} \tag{6.13}$$

where, for plane stress,

$$d_{11} = d_{22} = \frac{E}{1-v^2}$$

$$d_{21} = d_{12} = \frac{vE}{1-v^2}$$

$$d_{33} = \frac{E}{2(1+v)}$$

and for plane strain,

$$d_{11} = d_{22} = \frac{(1-v)E}{(1+v)(1-2v)}$$

$$d_{21} = d_{12} = \frac{vE}{(1+v)(1-2v)}$$

$$d_{33} = \frac{E}{2(1+v)}$$

$$\tag{6.14}$$

Step VI: Replace internal stresses $\{\sigma(x, y)\}$ with statically equivalent nodal forces $\{F^e\}$, relate nodal forces to nodal displacements $\{\delta^e\}$, and hence obtain element stiffness matrix $[K^e]$

The general equation that results from this step has been fully discussed in Chapter 3, step VI, and may be written as

$$\{F^e\} = [\int [B]^T [D][B] \, d(\text{vol})]\{\delta^e\} \qquad \text{(VI)}$$

the $[B]$ matrix being as defined in equation 6.12 and the $[D]$ matrix being as defined in equation 6.13.

Whereas, for the triangular element discussed in Chapter 5, both matrix $[B]$ and matrix $[D]$ contain only constant terms and can thus be taken outside the integral sign in equation VI, this is not true in the present case since $[B]$ involves terms in both x and y (see equations 6.9 and 6.12). Hence, in the evaluation of $[K^e]$ according to the equation

$$[K^e] = \int [B]^T [D][B] \, d(\text{vol}) \qquad (6.15)$$

which, for an element of constant thickness becomes

$$[K^e] = t \iint [B]^T [D][B] \, dx \, dy$$

the product $[B]^T [D][B]$ has to be evaluated first, and the terms of the resulting matrix have to be integrated over the area of the element. The calculation of $[K^e]$ is thus considerably more complicated than that indicated in Chapter 5 but still only involves standard matrix procedures. The final value of the $[K^e]$ matrix obtained from these calculations is given in equation 6.16, p. 88, where $p=a/b$. The reader is recommended to obtain one or two terms of this matrix in order to familiarise himself with the procedure.

Step VII: Establish stress–displacement matrix $[H]$

The required relationship is given in general equation V as

$$\{\sigma(x, y)\} = [D][B]\{\delta^e\} \qquad \text{(V)}$$

or

$$\{\sigma(x, y)\} = [H]\{\delta^e\} \qquad \text{(VII)}$$

where $[H]=[D][B]$ is the required stress–displacement matrix.

$$[K^e] = \frac{t}{12}
\begin{bmatrix}
4d_{11}p^{-1}+4d_{33}p & & & & & \\[4pt]
3d_{21}+3d_{33} & 4d_{22}p+4d_{33}p^{-1} & & & & \\[4pt]
2d_{11}p^{-1}-4d_{33}p & 3d_{21}-3d_{33} & 4d_{11}p^{-1}+4d_{33}p & & & \\[4pt]
-3d_{21}+3d_{33} & -4d_{22}p+2d_{33}p^{-1} & -3d_{21}-3d_{33} & 4d_{22}p+4d_{33}p^{-1} & & \\[4pt]
-4d_{11}p^{-1}+2d_{33}p & -3d_{21}+3d_{33} & -2d_{11}p^{-1}+2d_{33}p & 3d_{21}+3d_{33} & 4d_{11}p^{-1}+4d_{33}p & \\[4pt]
-3d_{21}+3d_{33} & 2d_{22}p-4d_{33}p^{-1} & 3d_{21}+3d_{33} & -4d_{22}p+2d_{33}p^{-1} & -3d_{21}+3d_{33} & 4d_{22}p+4d_{33}p^{-1}
\end{bmatrix}
\qquad \text{(symmetric)}$$

(6.16)

$$[H] = \frac{1}{ab}\begin{bmatrix} -d_{11}(b-y) & -d_{21}(a-x) & -d_{11}y & d_{21}(a-x) & d_{11}(b-y) & -d_{21}x & d_{11}y & d_{21}x \\ -d_{21}(b-y) & -d_{22}(a-x) & -d_{21}y & d_{22}(a-x) & d_{21}(b-y) & -d_{22}x & d_{21}y & d_{22}x \\ -d_{33}(a-x) & -d_{33}(b-y) & d_{33}(a-x) & -d_{33}y & -d_{33}x & d_{33}(b-y) & d_{33}x & d_{33}y \end{bmatrix}$$

$$(6.17)$$

Taking the $[D]$ matrix as defined in equation 6.13 and the $[B]$ matrix from equation 6.12, the product $[D][B]$ can be evaluated as shown in equation 6.17 (p. 89) where d_{11}, d_{22}, d_{12}, d_{21} and d_{33} have the values defined in equation 6.14 for plane stress and plane strain solutions.

Evaluation of equation VII using the value of $[H]$ defined in equation 6.17 gives the stresses $\{\sigma(x, y)\}$ at any point (x, y) on the element. The stresses thus obtained contain terms in x and y which the computer is unable to deal with. Consequently, in order to obtain the stresses at a specific point, the co-ordinates of that point are substituted into terms of the $[H]$ matrix. In this way, the stresses at each of the four nodes of the element can be determined. These may be represented by $\{\sigma^e\}$ where

$$\{\sigma^e\} = \begin{Bmatrix} \{\sigma(x_1, y_1)\} \\ \{\sigma(x_2, y_2)\} \\ \{\sigma(x_3, y_3)\} \\ \{\sigma(x_4, y_4)\} \end{Bmatrix} = \begin{bmatrix} [H(x_1, y_1)] \\ [H(x_2, y_2)] \\ [H(x_3, y_3)] \\ [H(x_4, y_4)] \end{bmatrix} \{\delta^e\} \qquad (6.18)$$

the value of $[H(x_1, y_1)]$, for example, being obtained simply by substituting the co-ordinates of node 1 into the expression for $[H]$ given in equation 6.17. A similar procedure is then followed for each of the other three nodes. Equation 6.18 may thus be rewritten as shown in equation 6.19 (p. 91). This equation may be summarised as

$$\{\sigma^e\} = [H^e]\{\delta^e\} \qquad (6.20)$$

thus defining the $[H^e]$ matrix for a rectangular element in a plane elasticity solution.

Since the strains vary linearly over this rectangular element, as discussed earlier, the stresses also vary linearly across the element for a constant value of x or y, in contrast to the uniform distribution of stresses over the triangular element discussed in Chapter 5. In general, this linear variation is not strictly correct and stresses calculated for adjacent elements are liable to be slightly discontinuous at the common nodes of the elements. The magnitudes of these discontinuities are, however, usually very small, especially if a fine mesh is employed, and a good approximation can be obtained by

$$
\{\sigma^e\} = \frac{1}{ab}
\left[
\begin{array}{cc:cc:cc:cc}
-d_{11}b & -d_{21}a & 0 & d_{21}a & d_{11}b & 0 & 0 & 0 \\
-d_{21}b & -d_{22}a & 0 & d_{22}a & d_{21}b & 0 & 0 & 0 \\
-d_{33}a & -d_{33}b & d_{33}a & 0 & 0 & d_{33}b & 0 & 0 \\
\hdashline
0 & -d_{21}a & -d_{11}b & d_{21}a & 0 & 0 & d_{11}b & 0 \\
0 & -d_{22}a & -d_{21}b & d_{22}a & 0 & 0 & d_{21}b & 0 \\
-d_{33}a & 0 & d_{33}a & -d_{33}b & 0 & 0 & 0 & d_{33}b \\
\hdashline
-d_{11}b & 0 & 0 & 0 & d_{11}b & -d_{21}a & 0 & d_{21}a \\
-d_{21}b & 0 & 0 & 0 & d_{21}b & -d_{22}a & 0 & d_{22}a \\
0 & -d_{33}b & 0 & 0 & -d_{33}a & d_{33}b & d_{33}a & 0 \\
\hdashline
0 & 0 & -d_{11}b & 0 & 0 & -d_{21}a & d_{11}b & d_{21}a \\
0 & 0 & -d_{21}b & 0 & 0 & -d_{22}a & d_{21}b & d_{22}a \\
0 & 0 & 0 & -d_{33}b & -d_{33}a & 0 & d_{33}a & d_{33}b
\end{array}
\right]
\underbrace{
\begin{array}{c}
\overbrace{}^{\substack{\text{node 1}\\ x=0\\ y=0}}
\overbrace{}^{\substack{\text{node 2}\\ x=0\\ y=b}}
\overbrace{}^{\substack{\text{node 3}\\ x=a\\ y=0}}
\overbrace{}^{\substack{\text{node 4}\\ x=a\\ y=b}}
\end{array}
}_{\{\delta^e\}}
\tag{6.19}
$$

taking an average of all the stress values calculated at a particular node.

Application

An example of the type of problem that can be solved by using the rectangular element described in this chapter is shown in Figure 6.3.

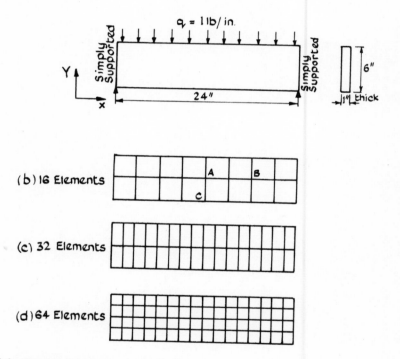

Fig. 6.3. Details of plane stress problem

This is a deep simply supported beam supporting a uniformly distributed load and it was analysed as a plane stress case. Three meshes were employed using 16, 32 and 64 elements respectively, as shown in Figure 6.3. Some of the results obtained are compared to an exact solution obtained from classical plane elasticity theory in Table 6.1.

Elements	Vertical deflection		Longitudinal stress
	Point A (in × 10⁻⁶)	Point B (in × 10⁻⁶)	Point C (lb/in²)
16	782	560	10·8
32	844	605	11·9
64	861	616	12·0
Exact solution	898	645	12·2

Table 6.1

It is clear from the table that as more elements are taken the accuracy of the finite element solution improves. The solution obtained using the 64-element mesh is in close agreement with the values given by the exact solution.

The stress discontinuities at the node points, which were mentioned earlier, are illustrated for this particular example in Figure 6.4. A substantial discontinuity is noted in the 16-element values at the mid-depth position but this is very much reduced in the 64-element case. In practice the discontinuity would be removed by averaging all the values calculated at a particular node.

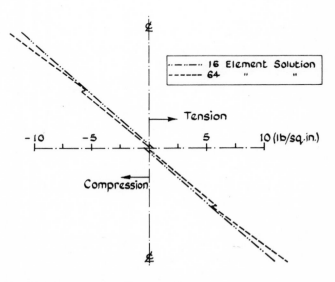

Fig. 6.4. Longitudinal stresses at midspan of beam

The rectangular plane elasticity element discussed in this chapter is slightly more accurate than the triangular element considered in Chapter 5, because it assumes a linear distribution of strain over the element and is thus better able to represent regions having a high stress gradient. Fewer rectangular elements would thus have to be taken to give results of equal accuracy to triangular elements. However, the triangular element has the advantage that it can be used for bodies with irregular boundary shapes and it is also more amenable to the production of graded meshes.

CHAPTER SEVEN

Rectangular Finite Element for Plate Flexure

In this chapter the application of the finite element technique to the solution of plate flexure problems is considered and the stiffness characteristics of a rectangular element are derived.

The thickness of the plate to be analysed is assumed to be small compared to its other dimensions and the deflection of the plate under load is assumed to be small compared to its thickness. These two assumptions are not particular to the finite element method and are also made in the classical solutions for plate flexure. These assumptions are necessary because if the thickness of a plate is large, then the plate has to be analysed as a three-dimensional problem and if the deflections under load are also large, then membrane forces are set up in the plane of the plate and these forces have to be taken into account in the analysis. The finite element method can be extended to the solution of these problems, but the technique for doing this is beyond the scope of the present chapter, which thus concentrates on the analysis of thin plates that undergo small deflections only. This is an extremely important class of problem having many practical applications such as in the design of flat-slab bridge decks and flooring units.

Derivation of rectangular element stiffness matrix

The derivation of the stiffness characteristics for the rectangular element once again follows the seven basic steps of Appendix1.

Step I: Choose suitable co-ordinate system and number nodes

The co-ordinate and node numbering systems defined for the rectangular element in plane elasticity solutions in Chapter 6 are also suitable for the plate flexure solutions. These systems are shown in Figure 7.1a, an additional z-axis now being taken normal to the

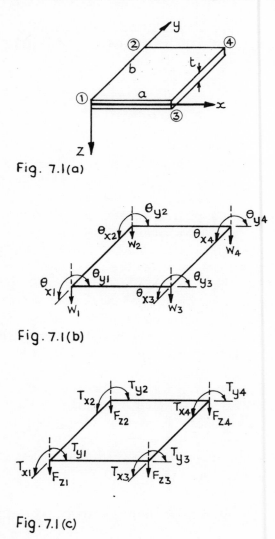

Fig. 7.1(a)

Fig. 7.1(b)

Fig. 7.1(c)

Fig. 7.1. *a* Co-ordinate system. *b* Nodal displacements. *c* Nodal forces

plane of the plate. In the plate flexure case, the element has three degrees of freedom at each node, namely two rotations and the transverse deflection. The lateral deflection is denoted by w, the rotation about the x-axis is denoted by θ_x and the rotation about the y-axis is denoted by θ_y. The positive directions of these rotations are defined according to the right-hand corkscrew rule. The element then has a total of twelve degrees of freedom as shown in Figure 7.1b. The corresponding moments and forces consist of two moments T_x and T_y and a shearing force F_z at each node, as shown in Figure 7.1c. The displacements at node 1 may be written as

$$\{\delta_1\} = \begin{Bmatrix} \theta_{x1} \\ \theta_{y1} \\ w_1 \end{Bmatrix}$$

and the corresponding moments and forces at node 1 may be written as

$$\{F_1\} = \begin{Bmatrix} T_{x1} \\ T_{y1} \\ F_{z1} \end{Bmatrix}$$

so that the complete displacement and force vectors for the element may be written as

$$\{\delta^e\} = \begin{Bmatrix} \{\delta_1\} \\ \{\delta_2\} \\ \{\delta_3\} \\ \{\delta_4\} \end{Bmatrix} = \begin{Bmatrix} \theta_{x1} \\ \theta_{y1} \\ w_1 \\ \hline \theta_{x2} \\ \theta_{y2} \\ w_2 \\ \hline \theta_{x3} \\ \theta_{y3} \\ w_3 \\ \hline \theta_{x4} \\ \theta_{y4} \\ w_4 \end{Bmatrix} \tag{7.1}$$

$$\{F^e\} = \begin{Bmatrix} \{F_1\} \\ \{F_2\} \\ \{F_3\} \\ \{F_4\} \end{Bmatrix} = \begin{Bmatrix} T_{x1} \\ T_{y1} \\ F_{z1} \\ \hdashline T_{x2} \\ T_{y2} \\ F_{z2} \\ \hdashline T_{x3} \\ T_{y3} \\ F_{z3} \\ \hdashline T_{x4} \\ T_{y4} \\ F_{z4} \end{Bmatrix} \qquad (7.2)$$

Since each of these vectors contains twelve terms, in this case the element stiffness matrix $[K^e]$ is a 12×12 matrix.

$$\{F^e\} = [K^e]\{\delta^e\} \qquad \text{(I)}$$

Step II: Choose displacement function $[f(x, y)]$ that defines displacement $\{\delta(x, y)\}$ at any point in element

In the case of plate flexure in which the deflections are small, the state of displacement at any point within the element may be represented by three components, i.e.

$$\{\delta(x, y)\} = \begin{Bmatrix} \theta_x \\ \theta_y \\ w \end{Bmatrix}$$

These are illustrated for a typical point A in Figure 7.2.

The two slopes θ_x and θ_y are related to the lateral displacement w by the expressions

$$\theta_x = -\partial w / \partial y \quad \text{and} \quad \theta_y = \partial w / \partial x$$

the positive directions of θ_x and θ_y being taken to coincide with those chosen for θ_{x1}, θ_{y1} etc. as illustrated earlier. Hence, once a displacement function is chosen for w, the functions for θ_x and θ_y are automatically defined.

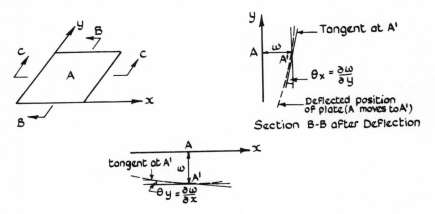

Fig. 7.2. Section c–c after deflection

Since the element has twelve degrees of freedom, twelve undetermined constants must be employed in the polynomial expression chosen to represent w. A suitable function is given in equation 7.3.

$$w = \alpha_1 + \alpha_2 x + \alpha_3 y + \alpha_4 x^2 + \alpha_5 xy + \alpha_6 y^2 + \alpha_7 x^3 + \alpha_8 x^2 y$$
$$+ \alpha_9 xy^2 + \alpha_{10} y^3 + \alpha_{11} x^3 y + \alpha_{12} xy^3 \quad (7.3)$$

It should be noted that when x or y is constant, this expression becomes the same as that taken for a beam element in equation 3.1. This displacement function gives the following expressions for the rotations.

$$\theta_x = -\frac{\partial w}{\partial y} = -(\alpha_3 + \alpha_5 x + 2\alpha_6 y + \alpha_8 x^2 + 2\alpha_9 xy + 3\alpha_{10} y^2$$
$$+ \alpha_{11} x^3 + 3\alpha_{12} xy^2)$$

and

$$\theta_y = \frac{\partial w}{\partial x} = \alpha_2 + 2\alpha_4 x + \alpha_5 y + 3\alpha_7 x^2 + 2\alpha_8 xy + \alpha_9 y^2$$
$$+ 3\alpha_{11} x^2 y + \alpha_{12} y^3$$

It is now necessary to check that this function ensures continuity of deflections and slopes in the solution. Consider one edge of the element, e.g. edge 1–2 where x is constant and equal to 0. The lateral displacement and slopes at any point on this edge are shown in Figure 7.3, and from equation 7.3 these are given by

$$w = \alpha_1 + \alpha_3 y + \alpha_6 y^2 + \alpha_{10} y^3$$
$$\theta_x = -(\alpha_3 + 2\alpha_6 y + 3\alpha_{10} y^2)$$
$$\theta_y = \alpha_2 + \alpha_5 y + \alpha_9 y^2 + \alpha_{12} y^3$$

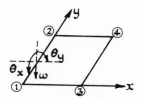

Fig. 7.3.

Considering the conditions at the ends of this edge, i.e. at nodes 1 and 2, when $y=0$, i.e. at node 1,

$$w = w_1 = \alpha_1$$
$$\theta_x = \theta_{x1} = -\alpha_3$$
$$\theta_y = \theta_{y1} = \alpha_2$$

and when $y=b$, i.e. at node 2,

$$w = w_2 = \alpha_1 + \alpha_3 b + \alpha_6 b^2 + \alpha_{10} b^3$$
$$\theta_x = \theta_{x2} = -(\alpha_3 + 2\alpha_6 b + 3\alpha_{10} b^2)$$
$$\theta_y = \theta_{y2} = \alpha_2 + \alpha_5 b + \alpha_9 b^2 + \alpha_{12} b^3$$

Thus only six equations are available to solve for eight unknown constants (α_1, α_2, α_3, α_5, α_6, α_9, α_{10}, α_{12}) and the constants cannot thus be determined. However, on close inspection it may be seen that w and θ_x contain the same four constants (α_1, α_3, α_6, α_{10}) whereas θ_y contains four different constants (α_2, α_5, α_9, α_{12}). Since four of the boundary condition equations refer to w and θ_x, a sufficient number of equations are available to solve for the constants related to these quantities, and w and θ_x can then be expressed in terms of the nodal displacements. The remaining two equations are not sufficient to determine the four unknown constants in θ_y. Hence, although it is clear that the lateral displacement w and rotation along the edge θ_x are completely defined by the end movements, θ_y, the rotation normal to the edge, is not uniquely specified. Since the end movements are made compatible during the solution, this means that continuity of w and θ_x are ensured along the edges where x is constant but θ_y, on the other hand, is discontinuous along this edge.

Only the specific case of edge 1–2 where $x=0$ has been considered above. However, the conclusion is generally true and along any edge a discontinuity of the normal slope can exist as shown in Figure 7.4. Because of this, the function chosen is not ideal and is called a 'non-conforming function'.

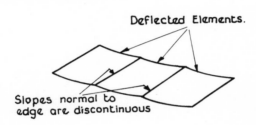

Deflected Elements.

Slopes normal to edge are discontinuous

Fig. 7.4.

Writing equation 7.3 in matrix form gives equation 7.4 (p. 102), which may be summarised as in general equation II as

$$\{\delta(x, y)\} = [f(x, y)]\{\alpha\} \qquad \text{(II)}$$

Step III: Express state of displacement $\{\delta(x, y)\}$ within element in terms of nodal displacements $\{\delta^e\}$

This process is undertaken in the same way as in Chapter 6, i.e. by substituting the nodal co-ordinate values into equation II and solving for $\{\alpha\}$. The substitution of the nodal co-ordinate values leads to the formation of the $[A]$ matrix defined in equation 6.7, i.e. $\{\delta^e\}=[A]\{\alpha\}$. Since in the present case, the element has a total of twelve degrees of freedom, $[A]$ is a 12×12 matrix and the terms are as given in equation 7.5, p. 103. This matrix now has to be inverted

$$
\begin{Bmatrix} \theta_x \\ \theta_y \\ w \end{Bmatrix} = \begin{bmatrix} 0 & 0 & -1 & 0 & -x & -2y & 0 & -x^2 & -2xy & -3y^2 & -x^3 & -3xy^2 \\ 0 & 1 & 0 & 2x & y & 0 & 3x^2 & 2xy & y^2 & 0 & 3x^2y & y^3 \\ 1 & x & y & x^2 & xy & y^2 & x^3 & x^2y & xy^2 & y^3 & x^3y & xy^3 \end{bmatrix} \begin{Bmatrix} \alpha_1 \\ \alpha_2 \\ \alpha_3 \\ \alpha_4 \\ \alpha_5 \\ \alpha_6 \\ \alpha_7 \\ \alpha_8 \\ \alpha_9 \\ \alpha_{10} \\ \alpha_{11} \\ \alpha_{12} \end{Bmatrix}
\tag{7.4}
$$

$$
[A] =
\begin{bmatrix}
1 & 0 & 0 & 0 & 0 & 0 & 0 & 0 & 0 & 0 & 0 & 0 \\
0 & 1 & 0 & 0 & 0 & 0 & 0 & 0 & 0 & 0 & 0 & 0 \\
0 & 0 & -1 & 0 & 0 & 0 & 0 & 0 & 0 & 0 & 0 & 0 \\
1 & 0 & b & 0 & 0 & b^2 & 0 & 0 & 0 & b^3 & 0 & 0 \\
0 & 1 & 0 & 0 & b & 0 & 0 & 0 & b^2 & 0 & 0 & b^3 \\
0 & 0 & -1 & 0 & 0 & -2b & 0 & 0 & 0 & -3b^2 & 0 & 0 \\
1 & a & 0 & a^2 & 0 & 0 & a^3 & 0 & 0 & 0 & 0 & 0 \\
0 & 1 & 0 & 2a & 0 & 0 & 3a^2 & 0 & 0 & 0 & 0 & 0 \\
0 & 0 & -1 & 0 & -a & 0 & 0 & -a^2 & 0 & 0 & -a^3 & 0 \\
1 & a & b & a^2 & ab & b^2 & a^3 & a^2b & ab^2 & b^3 & a^3b & ab^3 \\
0 & 1 & 0 & 2a & b & 0 & 3a^2 & 2ab & b^2 & 0 & 3a^2b & b^3 \\
0 & 0 & -1 & 0 & -a & -2b & 0 & -a^2 & -2ab & -3b^2 & -a^3 & -3ab^2
\end{bmatrix}
\qquad (7.5)
$$

where the rows are grouped by node:

- node 1: $x = 0$, $y = 0$
- node 2: $x = 0$, $y = b$
- node 3: $x = a$, $y = 0$
- node 4: $x = a$, $y = b$

and the required relationship is obtained from general equation III as

$$\{\delta(x, y)\} = [f(x, y)][A]^{-1}\{\delta^e\} \tag{III}$$

Step IV: Relate strains $\{\varepsilon(x, y)\}$ at any point to displacements $\{\delta(x, y)\}$ and hence to nodal displacements $\{\delta^e\}$

In the case of a plate flexure solution, the state of strain at any point may be represented by three components, namely the curvature in the x direction, the curvature in the y direction and the twist. The curvature in the x direction is equal to the rate of change of the slope in the x direction with respect to x and is

$$-\frac{\partial}{\partial x}\left(\frac{\partial w}{\partial x}\right) = -\frac{\partial^2 w}{\partial x^2}$$

Note that this expression for curvature corresponds to that given for the beam element in equation 3.5.

Similarly, the curvature in the y direction is

$$-\frac{\partial}{\partial y}\left(\frac{\partial w}{\partial y}\right) = -\frac{\partial^2 w}{\partial y^2}$$

Finally, the twist is equal to the rate of change of the slope in the x direction with respect to y and is

$$\frac{\partial}{\partial y}\left(\frac{\partial w}{\partial x}\right) = \frac{\partial^2 w}{\partial x \, \partial y}$$

Later, in Step 6, these curvatures and the twist are multiplied by the internal moments set up in the element in order to obtain the internal work done. The signs taken above correspond to the signs that are defined for the internal moments in Step V.

The internal moments M_x and M_y each act on two sides of the element. Similarly the twisting moments M_{xy} and M_{yx} each act on two sides, but since M_{xy} is equal to M_{yx}, one of these twists, say M_{xy}, can be considered to act on all four sides and this effect allowed for simply by doubling the twist term in the strain vector.

The state of 'strain' in the element can thus be represented as

$$\{\varepsilon(x, y)\} = \left\{\begin{array}{c} -\partial^2 w/\partial x^2 \\ -\partial^2 w/\partial y^2 \\ 2\partial^2 w/\partial x\,\partial y \end{array}\right\} \tag{7.6}$$

and substituting for w from equation 7.3 gives equation 7.7 (p. 106), which may be written as

$$\{\varepsilon(x, y)\} = [C]\{\alpha\} \tag{7.8}$$

so that the $[C]$ matrix has been defined for the case of plate flexure.

Proceeding as in Chapter 6 the required relationship between strains and nodal displacements is obtained as

$$\{\varepsilon(x, y)\} = [B]\{\delta^e\} \tag{IV}$$

$$[B] = [C][A]^{-1} \tag{7.9}$$

Once again, due to the size of the matrices involved, matrix $[B]$ will not be developed explicitly. In practice this could be done within the computer.

Step V: Relate internal stresses $\{\sigma(x, y)\}$ to strains $\{\varepsilon(x, y)\}$ and to nodal displacements $\{\delta^e\}$

In a plate flexure solution, the internal 'stresses' are really bending and twisting moments and the 'strains' are curvatures and twists as discussed in step IV. Thus for a plate flexure problem, the state of 'stress' can be represented by the three components M_x, M_y and M_{xy}, as indicated in equation 7.10.

$$\{\sigma(x, y)\} = \left\{\begin{array}{c} M_x \\ M_y \\ M_{xy} \end{array}\right\} \tag{7.10}$$

M_x and M_y are the internal bending moments per unit length and M_{xy} is the internal twisting moment per unit length set up within the element. If a small rectangular portion from within the finite element is considered, these internal bending and twisting moments

$$
\{\varepsilon(x,y)\} = \left\{ \begin{array}{c} -\partial^2 w/\partial x^2 \\[4pt] -\partial^2 w/\partial y^2 \\[4pt] 2\partial^2 w/\partial x\,\partial y \end{array} \right\} = \left\{ \begin{array}{c} -(2\alpha_4 + 6\alpha_7 x + 2\alpha_8 y + 6\alpha_{11} xy) \\[4pt] -(2\alpha_6 + 2\alpha_9 x + 6\alpha_{10} y + 6\alpha_{12} xy) \\[4pt] 2(\alpha_5 + 2\alpha_8 x + 2\alpha_9 y + 3\alpha_{11} x^2 + 3\alpha_{12} y^2) \end{array} \right\}
$$

$$
= \begin{bmatrix} 0 & 0 & 0 & -2 & 0 & 0 & -6x & -2y & 0 & 0 & -6xy & 0 \\ 0 & 0 & 0 & 0 & 0 & -2 & 0 & 0 & -2x & -6y & 0 & -6xy \\ 0 & 0 & 0 & 0 & 2 & 0 & 0 & 4x & 4y & 0 & 6x^2 & 6y^2 \end{bmatrix}
\left\{ \begin{array}{c} \alpha_1 \\ \alpha_2 \\ \alpha_3 \\ \alpha_4 \\ \alpha_5 \\ \alpha_6 \\ \alpha_7 \\ \alpha_8 \\ \alpha_9 \\ \alpha_{10} \\ \alpha_{11} \\ \alpha_{12} \end{array} \right\}
\qquad (7.7)
$$

may be shown acting as in Figure 7.5, where $M_{xy}=M_{yx}$. The 'stress–strain', i.e. the moment–curvature, relationships are known from plate bending theory to be as in equation (7.11).

$$
\left.
\begin{aligned}
M_x &= -\left(D_x \frac{\partial^2 w}{\partial x^2}+D_1 \frac{\partial^2 w}{\partial y^2}\right) \\[2mm]
M_y &= -\left(D_y \frac{\partial^2 w}{\partial y^2}+D_1 \frac{\partial^2 w}{\partial x^2}\right) \\[2mm]
M_{xy} &= 2D_{xy} \frac{\partial^2 w}{\partial x\,\partial y}
\end{aligned}
\right\}
\qquad (7.11)
$$

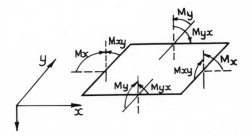

Fig. 7.5. Internal forces within elements

These relationships are written in general terms for an orthotropic plate, i.e. a plate which has different elastic properties in two perpendicular directions, a corrugated sheet being a typical example. D_x and D_y are the flexural rigidities in the x and y directions respectively, D_1 is a 'coupling' rigidity representing a Poisson's ratio type of effect, and D_{xy} is the torsional rigidity. On the other hand, an isotropic plate has the same elastic properties in all directions and for this particular case

$$
\begin{aligned}
D_x = D_y = D &= Et^3/12(1-v^2) \\
D_1 = vD \quad \text{and} \quad D_{xy} &= \tfrac{1}{2}(1-v)D
\end{aligned}
$$

Now equations 7.10 and 7.11 may be written in matrix form as

$$
\{\sigma(x,\,y)\} =
\begin{Bmatrix} M_x \\ M_y \\ M_{xy} \end{Bmatrix}
=
\begin{bmatrix}
D_x & D_1 & 0 \\
D_1 & D_y & 0 \\
0 & 0 & D_{xy}
\end{bmatrix}
\begin{Bmatrix}
-\partial^2 w/\partial x^2 \\
-\partial^2 w/\partial y^2 \\
2\partial^2 w/\partial x\,\partial y
\end{Bmatrix}
\qquad (7.12)
$$

that is,

$$\{\sigma(x, y)\} = [D]\{\varepsilon(x, y)\} \tag{7.13}$$

thus defining the $[D]$ matrix. Substituting for $\{\varepsilon(x, y)\}$ from general equation IV gives the required relationship between element stresses and the nodal displacements as usual, i.e.

$$\{\sigma(x, y)\} = [D][B]\{\delta^e\} \tag{V}$$

Step VI: Replace internal stresses $\{\sigma(x, y)\}$ with statically equivalent nodal forces $\{F^e\}$, relate nodal forces to nodal displacements $\{\delta^e\}$ and hence obtain element stiffness matrix $[K^e]$

This procedure is undertaken exactly as discussed in the previous chapters. Thus, the relationship between nodal loads $\{F^e\}$ and displacements $\{\delta^e\}$ is given in general equation VI as

$$\{F^e\} = [\int [B]^T [D][B] \, d(\text{vol})]\{\delta\} \tag{VI}$$

For the particular case of a plate bending element, the total work done by the internal moments and twists per unit length (M_x, M_y and M_{xy}) during a virtual displacement of the element is obtained by integrating the product of these moments and twists and their associated curvatures over the surface area of the element. Thus, the expression $\int^v d(\text{vol})$ in equation VI for the general case must be replaced by the expression $\int_0^b \int_0^a dx \, dy$ for the specific case of a rectangular plate bending element, so that equation VI must be rewritten as

$$\{F^e\} = [\int_0^b \int_0^a [B]^T [D][B] \, dx \, dy]\{\delta^e\}$$

and the element stiffness matrix $[K^e]$ is defined as

$$[K^e] = \int_0^b \int_0^a [B]^T [D][B] \, dx \, dy \tag{7.14}$$

Note that in the particular case of a one–dimensional beam element in Chapter 3 where the internal 'stress' was equivalent to the internal moment per unit length, $\int d(\text{vol})$ was replaced by $\int_0^l dx$. On the other hand, for the plane elasticity elements discussed in Chapters 5 and 6, the internal 'stresses' considered were the real stresses set up within the element, i.e. σ_x, σ_y and τ_{xy}, so that in these cases $\int d(\text{vol})$ was retained in this form.

The matrices required to calculate $[K^e]$ from equation 7.14 have all been derived in the present chapter for the plate flexure case. $[B]=[C][A]^{-1}$ from equation 7.9, where $[A]$ is defined in equation 7.5 and $[C]$ is defined in equation 7.7. The $[D]$ matrix has also been defined in equation 7.12. Once again, as in the case of the rectangular element in a plane elasticity solution, the matrix $[B]$ involves terms in x and y, and so the product of the matrices $[B^T]$ $[D][B]$ has to be integrated over the area of the element. The calculations are thus quite complex and are not now considered explicitly. However, the final value of the stiffness matrix $[K^e]$ obtained from these calculations is as given in equation 7.15 (p. 110). These expressions hold for the general orthotropic case. In any specific problem the appropriate values for the flexural rigidities D_x, D_y, D_1 and D_{xy} must be substituted.

Step VII: Establish stress–displacement matrix $[H]$

$$\{\sigma(x, y)\} = [H]\{\delta^e\} \tag{VII}$$

As usual $[H]$ is given by

$$[H] = [D][B] \tag{7.16}$$

Taking the $[D]$ matrix defined in equation 7.12 and the $[B]$ matrix from equation 7.9 the product $[D][B]$ can then be evaluated. Since the $[B]$ matrix contains terms in x and y, the calculated $[H]$ matrix also contains terms in x and y, and equation VII then gives the moments $\{\sigma(x, y)\}$ at any point (x, y) on the element. In order to obtain the moments at the four corners of the element, the four sets of nodal co-ordinates are simply substituted, as was done in the plane elasticity case in Chapter 6. The moments at the corners can then be expressed in terms of the nodal displacements as $\{\delta^e\} = [H^e]\{\delta^e\}$ where the value of $[H^e]$ is given in equation 7.17 (p. 112).

The manner in which the internal moments vary over the element is now considered. From equation 7.11,

$$M_x, M_y = f\left(\frac{\partial^2 w}{\partial x^2}, \frac{\partial^2 w}{\partial y^2}\right) \qquad M_{xy} = f\left(\frac{\partial^2 w}{\partial x \, \partial y}\right)$$

$$[K^e] = \frac{1}{15ab}
\begin{bmatrix}
SA \\
-SB & SC \\
-SD & SE & SF \\
SG & 0 & -SH & SA \\
0 & SI & SJ & SB & SC \\
-SH & SJ & SM & SD & SE & SF \\
SN & 0 & SO & SP & 0 & SQ & SA \\
0 & SR & -SS & 0 & ST & SU & -SB & SC \\
SO & -SS & SX & -SQ & -SU & SY & -SD & SE & SF \\
SP & 0 & -SQ & SN & 0 & -SO & SG & 0 & SH & SA \\
0 & ST & SU & 0 & SR & SS & 0 & SI & -SJ & -SB & SC \\
SQ & -SU & SY & SO & -SS & SX & -SH & -SJ & SM & SD & -SE & SF
\end{bmatrix}$$

symmetric

(7.15)

where

$$p = a/b$$
$$SA = 20a^2D_y + 8b^2D_{xy}$$
$$SB = 15abD_1$$
$$SC = 20b^2D_x + 8a^2D_{xy}$$
$$SD = 30apD_y + 15bD_1 + 6bD_{xy}$$
$$SE = 30bp^{-1}D_x + 15aD_1 + 6aD_{xy}$$
$$SF = 60p^{-2}D_x + 60p^2D_y + 30D_1 + 84D_{xy}$$
$$SG = 10a^2D_y - 2b^2D_{xy}$$
$$SH = -30apD_y - 6bD_{xy}$$
$$SI = 10b^2D_x - 8a^2D_{xy}$$
$$SJ = 15bp^{-1}D_x - 15aD_1 - 6aD_{xy}$$

$$SM = 30p^{-2}D_x - 60p^2D_y - 30D_1 - 84D_{xy}$$
$$SN = 10a^2D_y - 8b^2D_{xy}$$
$$SO = -15paD_y + 15bD_1 + 6bD_{xy}$$
$$SP = 5a^2D_y + 2b^2D_{xy}$$
$$SQ = 15apD_y - 6bD_{xy}$$
$$SR = 10b^2D_x - 2a^2D_{xy}$$
$$SS = 30bp^{-1}D_x + 6aD_{xy}$$
$$ST = 5b^2D_x + 2a^2D_{xy}$$
$$SU = 15bp^{-1}D_x - 6aD_{xy}$$
$$SX = 60p^{-2}D_x + 30p^2D_y - 30D_1 - 84D_{xy}$$
$$SY = -30p^{-2}D_x - 30p^2D_y + 30D_1 + 84D_{xy}$$

$$
[H^e] = \frac{1}{ab}
\begin{bmatrix}
-4aD_1 & -4aD_y & 2bD_{xy} & \vline & 2aD_1 & 2aD_y & 0 & \vline & 0 & 0 & 2bD_{xy} & \vline & 0 & 0 & 0 \\
4bD_x & 4bD_1 & -2aD_{xy} & \vline & -6pD_1 & -6pD_y & -2D_{xy} & \vline & -2bD_x & -2bD_1 & 0 & \vline & 0 & 0 & 0 \\
6p^{-1}D_x+6pD_1 & 6pD_y+6p^{-1}D_1 & -2D_{xy} & \vline & 4aD_1 & 4aD_y & 2bD_{xy} & \vline & 0 & 0 & -2D_{xy} & \vline & 0 & 0 & 0 \\
-2aD_1 & -2aD_y & 0 & \vline & 4bD_x & 4bD_1 & 2aD_{xy} & \vline & 0 & 0 & 0 & \vline & 0 & 0 & 2bD_{xy} \\
0 & 0 & -2bD_{xy} & \vline & 6p^{-1}D_x+6pD_1 & 6pD_y+6p^{-1}D_1 & 2D_{xy} & \vline & 0 & 0 & 2D_{xy} & \vline & 0 & 0 & 2D_{xy} \\
-6pD_1 & -6pD_y & 2D_{xy} & \vline & 0 & 0 & 0 & \vline & -4aD_1 & -4aD_y & -2bD_{xy} & \vline & 2aD_1 & 2aD_y & 0 \\
0 & 0 & -2bD_{xy} & \vline & 0 & 0 & 0 & \vline & -4bD_x & -4bD_1 & -2aD_{xy} & \vline & 0 & 0 & 0 \\
2bD_x & 2bD_1 & 0 & \vline & 0 & 0 & 2D_{xy} & \vline & 6p^{-1}D_x+6pD_1 & 6pD_y+6p^{-1}D_1 & 2D_{xy} & \vline & -6pD_1 & -6pD_y & 2D_{xy} \\
-6p^{-1}D_x & -6p^{-1}D_1 & 2D_{xy} & \vline & 0 & 0 & 0 & \vline & -2aD_1 & -2aD_y & 0 & \vline & 4bD_x & 4bD_1 & -2aD_{xy} \\
0 & 0 & 0 & \vline & 2bD_x & 2bD_1 & 0 & \vline & 0 & 0 & -2bD_{xy} & \vline & -4bD_x & -4bD_1 & -2bD_{xy} \\
0 & 0 & -2D_{xy} & \vline & -6p^{-1}D_x & -6p^{-1}D_1 & -2D_{xy} & \vline & -6pD_1 & -6pD_y & -2D_{xy} & \vline & 6p^{-1}D_x+6pD_1 & 6p^{-1}D_y+6p^{-1}D_1 & -2D_{xy} \\
0 & 0 & -2D_{xy} & \vline & 0 & 0 & 2D_{xy} & \vline & -6pD_1 & -6pD_y & -2D_{xy} & \vline & 2aD_{xy} & 2aD_{xy} & -2D_{xy}
\end{bmatrix}
\tag{7.17}
$$

and substituting for the curvatures and twists from equation 7.7,

$$M_x, M_y = f(2\alpha_4 + 6\alpha_7 x + 2\alpha_8 y + 6\alpha_{11} xy,$$

$$2\alpha_6 + 2\alpha_9 x + 6\alpha_{10} y + 6\alpha_{12} xy)$$

$$M_{xy} = f(\alpha_5 + 2\alpha_8 x + 2\alpha_9 y + 3\alpha_{11} x^2 + 3\alpha_{12} y^2)$$

so that, for any constant value of x or y, the bending moments M_x and M_y vary linearly across the element, just as do the stresses across the rectangular element in the plane elasticity solution. The twisting moment M_{xy}, on the other hand, is not linearly distributed but varies according to a quadratic expression in x when y is a constant and vice versa.

Once again, these assumed distributions are not strictly correct and discontinuities arise between elements as discussed in the plane elasticity solutions. Errors, however, are again small and good approximations can be obtained by averaging all the moment values calculated at a particular node.

All the required stiffness characteristics for a rectangular element in plate flexure have now been derived using once again the seven general steps.

Application

As an example of the type of problem that may be analysed using the element described in this chapter a square isotropic plate, clamped along all four edges and loaded by a uniformly distributed load q, is shown in Figures 7.6 and 7.7. The plate was divided into 4, 8, 16

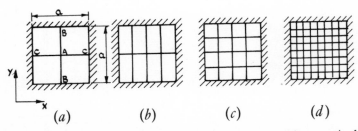

(a)   (b)   (c)   (d)

Fig. 7.6. Distribution of elements for clamped plate problem. *a* 4 elements. *b* 8 elements. *c* 16 elements. *d* 64 elements

Elements	Lateral deflection (w) at A	Bending moments $(M_x = M_y)$ at A	Bending moments $(M_{yB} = \mathbf{M}_{xC})$ at edge
4	0·00148	0·0462	−0·0355
8	0·00144	0·0363	−0·0418
16	0·00140	0·0278	−0·0476
64	0·00130	0·0240	−0·0503
Exact solution[1]	0·00126	0·0231	−0·0513
Multiplier	qa^4/D	qa^2	qa^2

Table 7.1

and 64 elements as shown and the results obtained from the finite element solutions are compared to results obtained from an exact solution[1] in Table 7.1. Excellent agreement is obtained between the 64-element solution and the exact solution, and it is apparent that the accuracy of the solution increases as the mesh is refined. The discontinuities that arise in the moment values are also shown in

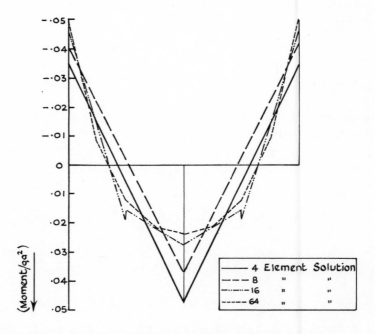

Fig. 7.7. Bending moments across central section of clamped plate

Figure 7.7. These are again seen to diminish as more elements are taken and have practically disappeared in the 64-element case.

Reference

1. TIMOSHENKO, S. P. and WOINOWSKY-KRIEGER, S. *Theory of plates and shells*. New York, McGraw-Hill Book Co., 1959.

CHAPTER EIGHT

Analysis of Folded-Plate, Box-Girder and Shell Structures using Rectangular Elements

Introduction

The rectangular plane elasticity element described in Chapter 6 and the rectangular plate bending element discussed in Chapter 7 can be combined to give a method of analysis for folded-plate and shell structures. Such structures are frequently used for roofing where their ability to span large areas without requiring intermediate supports makes them very attractive to the designer. Furthermore, a box-girder bridge is only a particular case of a folded-plate structure in which the plates are arranged so as to form a closed section.

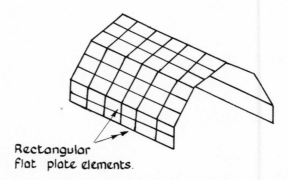

Rectangular
flat plate elements.

Fig. 8.1. Typical finite element idealisation of folded-plate structure

Thus the method described in this chapter can also be applied to the analysis of box-girder bridges.

A folded-plate structure, since it consists of a series of plane components, can be divided simply into a number of rectangular flat plate elements as shown in Figure 8.1. The surface of a continuously singly-curved shell can also be approximated by a series of rectangular flat plate elements as in Figure 8.2. The actual shell

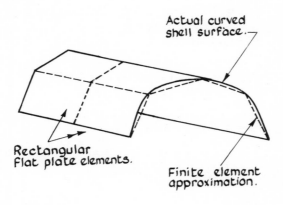

Fig. 8.2. Idealisation of curved surface using flat plate elements

structure is thus replaced by an 'equivalent' folded-plate structure and the accuracy of the approximation obviously depends on the number of elements taken. Clearly, more accurate solutions for continuously curved shell structures can be obtained by using finite elements which are themselves curved. However, such elements lie beyond the scope of this introductory text and quite satisfactory results can be obtained for many continuously curved shells by using the simple plane element discussed in this chapter.

Derivation of element stiffness matrix

Consider a folded-plate structure of the type shown in Figure 8.1. Now each individual plate of this structure and consequently each element on that plate, is generally subjected to load components acting both in its plane and normal to its plane, and exhibits both in-plane and flexural action. For small deflections these two actions are completely independent as far as the individual plate is concerned and the only interaction occurs at the ridges. Thus, as far as an element taken on the plate is concerned, its behaviour may be considered in two independent parts, i.e. an in-plane elasticity action and a plate bending action normal to its plane. Therefore it is assumed that flexural deflections and rotations of the element are only related to the forces normal to the plane and the in-plane displacements are only related to the in-plane forces.

Considering an element within a typical plate, from Chapter 6 it is known that the in-plane displacements and forces are as shown in Figure 8.3, and that these are related by the element in-plane stiffness

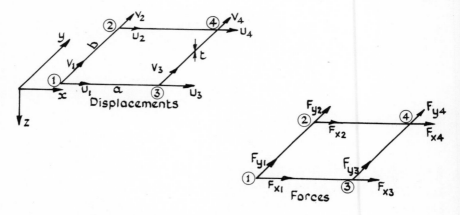

Fig. 8.3. In-plane nodal displacements and forces

matrix $[K^e]$, as defined in equation 6.16. Since the plates being considered at present are thin, the stress–strain relationships are those of plane stress, so that the appropriate values of the terms of the $[K^e]$ matrix can be obtained by substituting

$$d_{11} = d_{22} = \frac{E}{1-v^2}, \quad d_{21} = d_{12} = \frac{vE}{1-v^2} \quad \text{and} \quad d_{33} = \frac{E}{2(1+v)}$$

into the general expressions given in equation 6.16.

Considering now as an example the relationship between the in-plane forces and displacements at node 1 only, it is clear from Chapter 6 that

$$\begin{Bmatrix} F_{x1} \\ F_{y1} \end{Bmatrix} = \frac{t}{12} \begin{bmatrix} 4d_{11}p^{-1} & \\ \quad + & \text{symmetric} \\ 4d_{33}p & \\ \hline 3d_{21} & 4d_{22}p \\ \quad + & \quad + \\ 3d_{33} & 4d_{33}p^{-1} \end{bmatrix} \begin{Bmatrix} u_1 \\ v_1 \end{Bmatrix}$$

2×2 block in top left-hand corner of $[K^e]$ matrix defined in Equation 6.16.

and using suffix p to denote that in-plane behaviour is being referred to, this relationship may be summarised as

$$\{F_1{}^p\} = [K_{11}^p]\{\delta_1{}^p\} \tag{8.1}$$

where the subscripts 1, 1 show that $[K_{11}^p]$ relates the forces at node 1 to the displacements at node 1.

Also, from Chapter 7, it is known that the flexural displacements and forces of the element are as shown in Figure 8.4, and that these are related by the $[K^e]$ matrix defined in equation 7.15. Considering

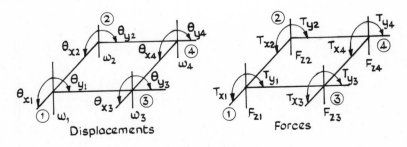

Fig. 8.4. Nodal displacements and forces normal to plane

once again the relationship between the forces and displacements at node 1 only, it is known from Chapter 7 that

$$\begin{Bmatrix} T_{x1} \\ T_{y1} \\ F_{z1} \end{Bmatrix} = \frac{1}{15ab} \begin{bmatrix} SA & \vdots & \text{symmetric} \\ \hline -SB & SC & \vdots \\ \hline -SD & \vdots & SE & \vdots & SF \end{bmatrix} \begin{Bmatrix} \theta_{x1} \\ \theta_{y1} \\ w_1 \end{Bmatrix}$$

3×3 block in top left-hand corner of $[K^e]$ matrix defined in equation 7.15.

and using suffix b to denote that bending normal to the plane is being referred to, this relationship may be summarised as

$$\{F_1{}^b\} = [K_{11}^b]\{\delta_1{}^b\} \tag{8.2}$$

In addition to the two in-plane displacements, the one lateral displacement and the two lateral rotations considered so far, the element can also undergo an in-plane 'drilling' or 'twisting' rotation of the type shown in Figure 8.5. Denoting this in-plane rotation by θ_z and the moment associated with it by T_z, the additional displacements and forces shown in Figure 8.6 must be considered for the

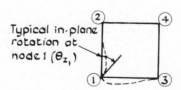

Fig. 8.5.

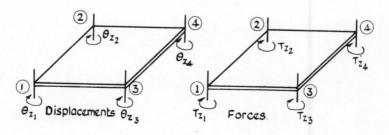

Fig. 8.6. Nodal 'drilling' or 'twisting' displacements and forces

element. Now the plate is extremely stiff in its plane so that this in-plane rotation of the element is very small and can be neglected. The only reason for introducing it at the present stage is that it makes subsequent operations of assembling the elements very much more convenient, as will become apparent later. In order to include the θ_z term, which is known to be negligible and not to affect the other terms in the element stiffness matrix, the relationship between the θ_z and T_z quantities may be expressed by 'dummy' equations as follows.

For example, at node 1, $\qquad T_{z1} = 0.\theta_{z1}$ \hfill (8.3)

Six displacement and six force components are thus considered at each node of the element. Those at node 1, for example, are shown in Figure 8.7. The positive directions of the rotations and moments are once again given by the right-hand corkscrew rule.

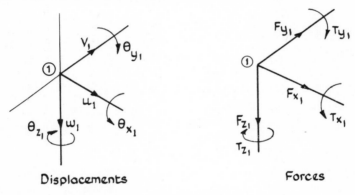

Displacements   Forces

Fig. 8.7. Total displacements and forces at node 1

All the displacements at node 1 may be written as a vector as shown in equation 8.4,

$$\{\delta_1\} = \begin{Bmatrix} \{\delta_1{}^p\} \\ \{\delta_1{}^b\} \\ \theta_{z1} \end{Bmatrix} = \begin{Bmatrix} u_1 \\ v_1 \\ \hline \theta_{x1} \\ \theta_{y1} \\ w_1 \\ \hline \theta_{z1} \end{Bmatrix} \qquad (8.4)$$

and the forces at node 1 may similarly be represented by a vector as shown in equation 8.5,

$$\{F_1\} = \begin{Bmatrix} \{F_1{}^p\} \\ \\ \{F_1{}^b\} \\ \\ T_{z1} \end{Bmatrix} = \begin{Bmatrix} F_{x1} \\ F_{y1} \\ \hline T_{x1} \\ T_{y1} \\ F_{z1} \\ \hline T_{z1} \end{Bmatrix} . \tag{8.5}$$

Using the relationships obtained in equations 8.1, 8.2 and 8.3, the relationship between all these forces and displacements at node 1 can be expressed as

$$\{F_1\} = \begin{Bmatrix} \{F_1{}^p\} \\ \{F_1{}^b\} \\ T_{z1} \end{Bmatrix} = \begin{bmatrix} [K_{11}^p] & 0 & 0 \\ 0 & [K_{11}^b] & 0 \\ 0 & 0 & 0 \end{bmatrix} \begin{Bmatrix} \{\delta_1{}^p\} \\ \{\delta_1{}^b\} \\ \theta_{z1} \end{Bmatrix} \tag{8.6}$$

where, since $[K_{11}^p]$ is a 2×2 matrix and $[K_{11}^b]$ is a 3×3 matrix, the complete linking matrix in equation 8.6 is a 6×6 matrix.

Now, so far only the relationship between the forces and displacements at node 1 has been considered, as being a typical example. Similar relationships of course exist between the forces and displacements at the other three nodes of the element. Furthermore, a similar relationship also exists between the forces at one node and the displacements at another node. For example, the forces at node 2 may be expressed in terms of the displacements at node 3 as follows.

$$\{F_2\} = \begin{Bmatrix} \{F_2{}^p\} \\ \{F_2{}^b\} \\ T_{z2} \end{Bmatrix} = \begin{bmatrix} [K_{23}^p] & 0 & 0 \\ 0 & [K_{23}^b] & 0 \\ 0 & 0 & 0 \end{bmatrix} \begin{Bmatrix} \{\delta_3{}^p\} \\ \{\delta_3{}^b\} \\ \theta_{z3} \end{Bmatrix} \tag{8.7}$$

Gathering all these relationships together the force–displacement relationship for the complete element may be written as shown in equation 8.8 (p. 123).

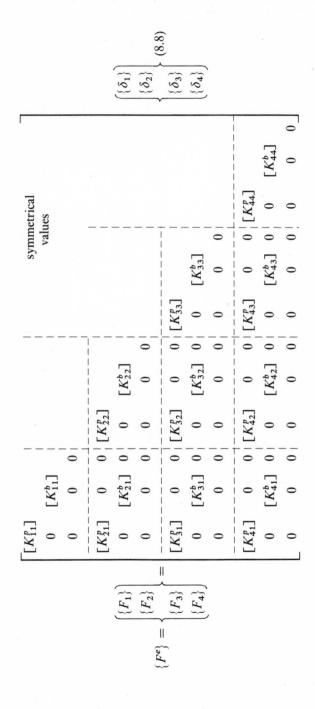

$$\{F^e\} = \begin{Bmatrix} \{F_1\} \\ \{F_2\} \\ \{F_3\} \\ \{F_4\} \end{Bmatrix} = \begin{bmatrix} [K_{11}^p] & 0 & 0 & [K_{22}^p] & 0 & 0 & [K_{33}^p] & 0 & 0 & [K_{44}^p] & 0 & 0 \\ 0 & [K_{11}^b] & 0 & & & & & & & & & \\ 0 & 0 & 0 & & & & & & & & & \\ [K_{21}^p] & 0 & 0 & [K_{22}^p] & 0 & 0 & & & & & & \\ 0 & [K_{21}^b] & 0 & 0 & [K_{22}^b] & 0 & & & & & & \\ 0 & 0 & 0 & 0 & 0 & 0 & & & & & & \\ [K_{31}^p] & 0 & 0 & [K_{32}^p] & 0 & 0 & [K_{33}^p] & 0 & 0 & & & \\ 0 & [K_{31}^b] & 0 & 0 & [K_{32}^b] & 0 & 0 & [K_{33}^b] & 0 & & & \\ 0 & 0 & 0 & 0 & 0 & 0 & 0 & 0 & 0 & & & \\ [K_{41}^p] & 0 & 0 & [K_{42}^p] & 0 & 0 & [K_{43}^p] & 0 & 0 & [K_{44}^p] & 0 & 0 \\ 0 & [K_{41}^b] & 0 & 0 & [K_{42}^b] & 0 & 0 & [K_{43}^b] & 0 & 0 & [K_{44}^b] & 0 \\ 0 & 0 & 0 & 0 & 0 & 0 & 0 & 0 & 0 & 0 & 0 & 0 \end{bmatrix} \begin{Bmatrix} \{\delta_1\} \\ \{\delta_2\} \\ \{\delta_3\} \\ \{\delta_4\} \end{Bmatrix} \quad (8.8)$$

symmetrical
values

The element stiffness matrix $[K^e]$, which, in this case, is a 24×24 matrix, is thus defined. The stiffness matrix for the 'folded-plate' element has thus been established and it is seen that all the terms of this stiffness matrix have been derived previously in either the plane stress or plate bending solutions of Chapters 6 and 7. This stiffness matrix relates the forces and displacements of the element in its own particular co-ordinate system, known as the 'local co-ordinate system' and shown in Figure 8.8. Now in the plane elasticity and plate

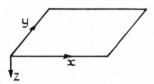

Fig. 8.8. Local co-ordinate system

flexure problems considered in Chapters 6 and 7, all the elements taken were co-planar. However, in folded-plate and shell problems, just as in the case of the framework problems discussed in Chapter 2 adjacent elements are normally inclined to one another. Consequently, before the elements can be assembled and the stiffness matrix of the complete structure formed, the displacements and forces of the individual elements must all be expressed in a common co-ordinate system. This system is referred to as either the 'fixed' or the 'global' co-ordinate system and is denoted by $\bar{x}$, $\bar{y}$ and $\bar{z}$, the displacements in this system, which are six in number, being called $\bar{u}$, $\bar{v}$, $\bar{\theta}_x$, $\bar{\theta}_y$, $\bar{w}$ and $\bar{\theta}_z$, as shown in Figure 8.9. For most structures, it is usually

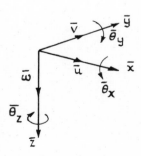

Fig. 8.9. Global co-ordinate system

convenient to take the $\bar{x}$ and $\bar{y}$ axes in the horizontal plane with $\bar{z}$ as the vertical axis, as shown in Figure 8.10.

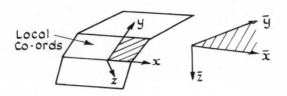

Fig. 8.10. Typical local and global co-ordinate system

Considering a typical element and denoting the angle between the x-axis and the $\bar{x}$-axis as $\phi x\bar{x}$, that between the x-axis and the $\bar{y}$-axis as $\phi x\bar{y}$, etc., the relationship between the displacements in the local and global systems at node 1 of this element may be written as

$$
\begin{Bmatrix} u_1 \\ v_1 \\ \theta_{x1} \\ \theta_{y1} \\ w_1 \\ \theta_{z1} \end{Bmatrix} =
\begin{bmatrix}
\cos\phi_{x\bar{x}} & \cos\phi_{x\bar{y}} & 0 & 0 & \cos\phi_{x\bar{z}} & 0 \\
\cos\phi_{y\bar{x}} & \cos\phi_{y\bar{y}} & 0 & 0 & \cos\phi_{y\bar{z}} & 0 \\
0 & 0 & \cos\phi_{x\bar{x}} & \cos\phi_{x\bar{y}} & 0 & \cos\phi_{x\bar{z}} \\
0 & 0 & \cos\phi_{y\bar{x}} & \cos\phi_{y\bar{y}} & 0 & \cos\phi_{y\bar{z}} \\
\cos\phi_{z\bar{x}} & \cos\phi_{z\bar{y}} & 0 & 0 & \cos\phi_{z\bar{z}} & 0 \\
0 & 0 & \cos\phi_{z\bar{x}} & \cos\phi_{z\bar{y}} & 0 & \cos\phi_{z\bar{z}}
\end{bmatrix}
\begin{Bmatrix} \bar{u}_1 \\ \bar{v}_1 \\ \bar{\theta}_{x1} \\ \bar{\theta}_{y1} \\ \bar{w}_1 \\ \bar{\theta}_{z1} \end{Bmatrix}
\quad (8.9)
$$

This equation may be summarised as

$$\{\delta_1\} = [\bar{T}]\{\bar{\delta}_1\} \qquad (8.10)$$

The reasons for including the θ_z term in the element stiffness matrix in local co-ordinates is now apparent. Because θ_z was included, the $[\bar{T}]$ matrix in equation 8.10 is now a square symmetrical matrix, so that subsequent computer operations involving this matrix become very much more convenient.

The inclinations between the local and global co-ordinate systems are exactly the same at the other three nodes of the element as they are at node 1, so that relationships similar to those of equations

8.9 and 8.10 also exist at these other nodes. Thus, for the complete element,

$$\begin{Bmatrix} \{\delta_1\} \\ \{\delta_2\} \\ \{\delta_3\} \\ \{\delta_4\} \end{Bmatrix} = \begin{bmatrix} [\overline{T}] & 0 & 0 & 0 \\ 0 & [\overline{T}] & 0 & 0 \\ 0 & 0 & [\overline{T}] & 0 \\ 0 & 0 & 0 & [\overline{T}] \end{bmatrix} \begin{Bmatrix} \{\bar{\delta}_1\} \\ \{\bar{\delta}_2\} \\ \{\bar{\delta}_3\} \\ \{\bar{\delta}_4\} \end{Bmatrix} \qquad (8.11)$$

that is,

$$\{\delta^e\} = [T]\{\bar{\delta}^e\} \qquad (8.12)$$

$[T]$ is known as the co-ordinate transformation matrix for the element. It is a 24×24 symmetrical matrix composed of 6×6 sub-matrices arranged along the leading diagonal.

Having thus defined the transformation matrix, the required element stiffness matrix in the fixed co-ordinate system $[\overline{K}^e]$ can now be evaluated according to the relationship given in equation 2.15, i.e. $[\overline{K}^e] = [T]^T [K^e][T]$. Having obtained the element stiffness matrix in the global co-ordinate system, subsequent operations leading to the solution of the problem are then similar to those considered previously.

Once the nodal displacements have been calculated in the global co-ordinate system they are transformed back to the local co-ordinate systems and the internal in-plane stresses of any element can then be obtained simply by multiplying the in-plane displacements by the stress–displacement matrix $[H]$ given in equation 6.17. Similarly, the internal bending and twisting moments of the element can be obtained by multiplying the flexural displacements by the $[H]$ matrix defined in equation 7.17.

Co-planar elements

Before concluding this discussion of the analysis of folded–plate and shell structures, one possible source of difficulty must be mentioned. So far it has been assumed that all the elements taken across a section of such a structure are inclined to each other. This is indeed true when a curved shell is analysed as an equivalent folded plate. However, very often in the analysis of folded plates it is found that the dimensions are such that several elements have to be taken across the width of each plate in order to achieve a satisfactory degree of accuracy (Figure 8.11). Such a mesh consists of a combination of co-planar and inclined elements.

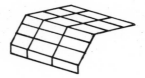

Fig. 8.11. Typical finite element idealisation

Now the element stiffness matrix in local co-ordinates $[K^e]$, as defined in equation 8.8, was established with respect to five components of force and displacement at each node (θ_z and T_z being dummy quantities at that stage). However, in the global co-ordinate system, six force and displacement components are considered at each node. In other words, during the assembly of the elements, six compatibility conditions are satisfied and six equilibrium equations are set up at each node. Thus at the junction of two inclined elements, six equations in the fixed co-ordinate system are formed from ten independent equations in the local co-ordinate system and this is satisfactory. However, at the junction of two co-planar elements, although each element contributes five equations these equations are similar since the co-ordinate directions of the elements coincide. Thus, when six equations in the fixed co-ordinate system are set up these are in fact formed from only five independent equations in the local co-ordinate system and two of the six equations so formed are dependent. The overall stiffness matrix of the structure is then singular and cannot be inverted to give the required solution.

In order to overcome this difficulty, a system of localised co-ordinate transformation has to be introduced. At the junction of two inclined elements, equilibrium is established with respect to six force components in the global co-ordinate system in the normal way. At the junction of two co-planar elements, equilibrium is only established with respect to the five force components in local co-ordinates that have been assumed to act on the elements. For example, the $\bar{y}$ and $\bar{z}$ axes taken at a typical cross-section could be as shown in

Figure 8.12. This process of localised transformation can be carried out automatically by the computer. The machine can check the location of each element relative to the ridges of the structure before forming the co-ordinate transformation matrix for that particular element.

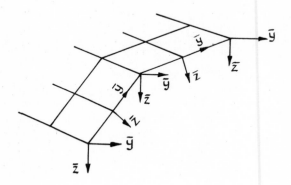

Fig. 8.12.

Since the solution of a folded-plate problem involves setting up six equilibrium equations at each node, more computer storage space and time is required than for any previous solution considered. The use of special programming techniques to achieve economy in computer time and storage, such as those that are discussed in Chapter 10, is thus essential in obtaining many folded-plate solutions.

Application

A typical example of the type of structure that can be analysed using the method described in this chapter is shown in Figure 8.13. This is a two-bay folded-plate structure subjected to a uniformly distributed load of 0.25 lb/in^2 acting normal to the plane of each of the inclined plates. The structure is supported at its ends by diaphragms that are considered to be infinitely rigid in their own planes and perfectly flexible in directions normal to these planes. Since the structure is symmetrical about its longitudinal and transverse centre-lines, only the elements taken on one-quarter of the structure

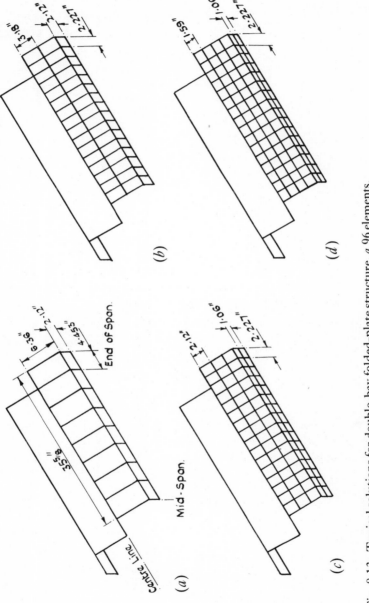

Fig. 8.13. Typical solutions for double-bay folded-plate structure. *a* 96 elements. *b* 320 elements. *c* 512 elements. *d* 640 elements

are shown in Figure 8.13 and the different meshes employed are seen to involve 96, 320, 512 and 640 elements respectively. In Table 8.1, the values obtained from these solutions for the vertical deflections of the ridges at midspan are compared to results obtained from

Elements	Vertical deflections (in)		
	Outer ridge	Top ridge	Centre ridge
96	0·08851	0·01134	0·01981
320	0·07381	0·01691	0·02598
512	0·07477	0·01811	0·02747
640	0·07373	0·01860	0·02802
Exact solution[1]	0·07364	0·01922	0·02876

Table 8.1

an accurate solution.[1] The agreement between the fine-mesh solution and the exact solution is seen to be excellent. Although only a few values are compared in Table 8.1 it must be appreciated that the finite element method gives a comprehensive picture of all the stresses, moments and deflections set up within the structure.

Reference

1. GOLDBERG, J. E. and LEVE, H. L. Theory of prismatic folded plate structures. *International Association for Bridge and Structural Engineering*, Vol. 17, 1959, pp. 59–86.

CHAPTER NINE

Axially Symmetric Continua

Introduction

There are many problems of stress analysis of very considerable practical importance concerned with axially symmetric structures such as pressure vessels, rotating discs, rocket nozzles and cases, cooling towers, column bases on soil, machine parts such as pistons, roof domes and storage tanks, and so on, where both the shape and the applied loads are axially symmetric. Using elements of revolution, such as those shown in Figure 9.1, these problems can be analysed readily using the finite element method. Such axisymmetric problems can be separated into two main categories—shells of revolution in which the thickness of the structure is small compared with its diameter, and solid bodies and bodies of revolution where the structure is thick compared with its diameter. These categories are now discussed in turn.

Axisymmetric shells

An axisymmetric shell structure can be idealised by a series of conical frustum-shaped elements as shown in Figure 9.1. This element was first suggested by Grafton and Strome[1] and later its use was extended by Percy *et al.*[2] It may be noted that this single element has two ring nodes. Since both in-plane and out of plane displacements and forces have to be considered in shell structures, the displacement vector for each node contains axial and radial movements as well as a rotation (Figure 9.1).

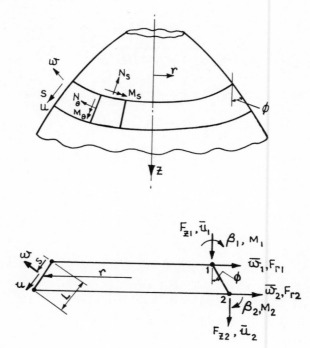

Fig. 9.1. Axisymmetric shell structure having two ring nodes

Step I: Choose suitable co-ordinate system and number nodes

It is convenient to use cylindrical polar co-ordinates (r, z) and considering node 1 in Figure 9.1 the nodal displacement vector is written as

$$\{\delta_1\} = \left\{ \begin{matrix} \bar{u}_1 \\ \bar{w}_1 \\ \beta_1 \end{matrix} \right\}$$

where $\bar{u}$ is the axial displacement, and $\bar{w}$ is the radial displacement, in global co-ordinates, and β is the rotation. The corresponding forces at node 1 are

$$\{F_1\} = \begin{Bmatrix} F_{z1} \\ F_{r1} \\ M_1 \end{Bmatrix}$$

where F_z is the axial force, and F_r is the radial force in global co-ordinates and M is the meridional moment.

The complete displacement and force vectors for the element, i.e. nodes 1 and 2, may then be written as

$$\{\delta^e\} = \begin{Bmatrix} \{\delta_1\} \\ \{\delta_2\} \end{Bmatrix} = \begin{Bmatrix} \bar{u}_1 \\ \bar{w}_1 \\ \beta_1 \\ \bar{u}_2 \\ \bar{w}_2 \\ \beta_2 \end{Bmatrix} \tag{9.1}$$

$$\{F^e\} = \begin{Bmatrix} \{F_1\} \\ \{F_2\} \end{Bmatrix} = \begin{Bmatrix} F_{z1} \\ F_{r1} \\ M_1 \\ F_{z2} \\ F_{r2} \\ M_2 \end{Bmatrix} \tag{9.2}$$

Thus each axisymmetric shell element has six degrees of freedom and the complete element stiffness matrix $[K^e]$ is of size 6×6.

$$\{F^e\} = [K^e]\{\delta^e\} \tag{I}$$

Step II: Choose displacement function $[f(r, z)]$ that defines displacement $\{\delta(r, z)\}$ at any point in element

Because of the inclination of the element ϕ to the z-axis of the shell, it is more convenient to specify the element displacement function in terms of local element co-ordinates (r, s), where u is the in-plane displacement and w is the displacement normal to the plane (Figure 9.1). Since there are six degrees of freedom per element, six unknown coefficients have to be taken in the polynomials representing the permitted displacement pattern. Equation 9.3 gives a suitable set of relationships in which the in-plane displacement function u varies linearly in s and w varies as a cubic in s.

$$\left.\begin{aligned}
u &= \alpha_1 + \alpha_2 s \\
w &= \alpha_3 + \alpha_4 s + \alpha_5 s^2 + \alpha_6 s^3 \\
\frac{dw}{ds} &= \alpha_4 + 2\alpha_5 s + 3\alpha_6 s^2
\end{aligned}\right\} \tag{9.3}$$

It may be noted that the third equation for the rotation dw/ds is obtained by differentiating the second. Writing equation 9.3 in matrix form, the relationship in equation 9.4 between the element displacements u and w and the undetermined coefficient α is obtained.

$$\begin{Bmatrix} u \\ w \\ \dfrac{dw}{ds} \end{Bmatrix} = \begin{bmatrix} 1 & s & 0 & 0 & 0 & 0 \\ 0 & 0 & 1 & s & s^2 & s^3 \\ 0 & 0 & 0 & 1 & 2s & 3s^2 \end{bmatrix} \begin{Bmatrix} \alpha_1 \\ \alpha_2 \\ \alpha_3 \\ \alpha_4 \\ \alpha_5 \\ \alpha_6 \end{Bmatrix} \tag{9.4}$$

Equation 9.4 may be written in matrix form as shown in equation II.

$$\{\delta(r, s)\} = [f(r, s)]\{\alpha\} \tag{II}$$

Step III: Express state of displacement $\{\delta(r, s)\}$ within element in terms of nodal displacements $\{\delta^e\}$

This step is achieved by substituting the values of the nodal co-ordinates (r, s) into equation II and solving to obtain a relationship for the coefficients $\{\alpha\}$. For example, at node 1, $s=0$, and thus from equation 9.3,

$$u_1 = \alpha_1, \qquad w_1 = \alpha_3, \quad \text{and} \quad \left(\frac{dw}{ds}\right)_1 = \alpha_4 \qquad (9.5a)$$

At node 2, $s=L$, and thus

$$u_2 = \alpha_1 + \alpha_2 L,$$

$$w_2 = \alpha_3 + \alpha_4 L + \alpha_5 L^2 + \alpha_6 L^3 \qquad (9.5b)$$

and

$$\left(\frac{dw}{ds}\right)_2 = \alpha_4 + 2\alpha_5 L + 3\alpha_6 L^2$$

Expressing equations 9.5a and 9.5b in matrix form,

$$\begin{Bmatrix} u_1 \\ w_1 \\ \left(\dfrac{dw}{ds}\right)_1 \\ u_2 \\ w_2 \\ \left(\dfrac{dw}{ds}\right)_2 \end{Bmatrix} = \begin{bmatrix} 1 & 0 & 0 & 0 & 0 & 0 \\ 0 & 0 & 1 & 0 & 0 & 0 \\ 0 & 0 & 0 & 1 & 0 & 0 \\ 1 & L & 0 & 0 & 0 & 0 \\ 0 & 0 & 1 & L & L^2 & L^3 \\ 0 & 0 & 0 & 1 & 2L & 3L^2 \end{bmatrix} \begin{Bmatrix} \alpha_1 \\ \alpha_2 \\ \alpha_3 \\ \alpha_4 \\ \alpha_5 \\ \alpha_6 \end{Bmatrix} \qquad (9.6)$$

To solve for $\{\alpha\}$ in terms of the nodal displacements, the 6×6 matrix of equation 9.6 could simply be inverted. However, in this case the solution for $\{\alpha\}$ can easily be obtained directly from equations 9.5a and 9.5b. Equations 9.5a give values for α_1, α_3 and α_4 in terms of the local deformations of node 1. Using these in equations 9.5b the remaining values, α_2, α_5 and α_6 are found to be

$$\alpha_2 = (u_2 - u_1)/L$$

$$\alpha_5 = -\frac{2}{L}\left(\frac{dw}{ds}\right)_1 - \frac{1}{L}\left(\frac{dw}{ds}\right)_2 + \frac{3}{L}(w_2 - w_1)$$

$$\alpha_6 = \frac{1}{L^2}\left(\frac{dw}{ds}\right)_2 + \frac{1}{L^2}\left(\frac{dw}{ds}\right)_1 - \frac{2}{L^3}(w_2 - w_1)$$

Writing these in matrix form the desired relationship given in equation 9.7 is obtained.

$$
\begin{Bmatrix} \alpha_1 \\ \alpha_2 \\ \alpha_3 \\ \alpha_4 \\ \alpha_5 \\ \alpha_6 \end{Bmatrix} =
\begin{bmatrix}
1 & 0 & 0 & 0 & 0 & 0 \\
-\dfrac{1}{L} & 0 & 0 & \dfrac{1}{L} & 0 & 0 \\
0 & 1 & 0 & 0 & 0 & 0 \\
0 & 0 & 1 & 0 & 0 & 0 \\
0 & -\dfrac{3}{L^2} & -\dfrac{2}{L} & 0 & \dfrac{3}{L^2} & -\dfrac{1}{L} \\
0 & \dfrac{2}{L^3} & \dfrac{1}{L^2} & 0 & -\dfrac{2}{L^3} & \dfrac{1}{L^2}
\end{bmatrix}
\begin{Bmatrix} u_1 \\ w_1 \\ \left(\dfrac{dw}{ds}\right)_1 \\ u_2 \\ w_2 \\ \left(\dfrac{dw}{ds}\right)_2 \end{Bmatrix}
\tag{9.7}
$$

The reader can check that this is correct by multiplying the 6×6 matrix of equation 9.7 by that in equation 9.6 to obtain the identity matrix.

Now the displacements $\begin{Bmatrix} u \\ w \end{Bmatrix}$ at any point within the element are expressed in terms of the coefficients $\{\alpha\}$ by equation 9.4, as

$$
\begin{Bmatrix} u \\ w \end{Bmatrix} =
\begin{bmatrix}
1 & s & 0 & 0 & 0 & 0 \\
0 & 0 & 1 & s & s^2 & s^3
\end{bmatrix}
\begin{Bmatrix} \alpha_1 \\ \alpha_2 \\ \alpha_3 \\ \alpha_4 \\ \alpha_5 \\ \alpha_6 \end{Bmatrix}
\tag{9.8}
$$

Thus on combining equations 9.7 and 9.8 the displacements u and w can be expressed at any point within the element in terms of the nodal displacement u_1, w_1, $(dw/ds)_1$, u_2, w_2 and $(dw/ds)_2$. Ordinarily this completes step III. However, these nodal displacements are in terms of the local (element) co-ordinates (r, s) and must now be transformed to the global co-ordinates (r, z) and the global deformations $\bar{u}$, $\bar{w}$ and β, since the geometry and loading of the structure is specified in these global co-ordinates.

Referring to Figure 9.2 it is seen that

$$u_i = \bar{u}_i \cos \phi + \bar{w}_i \sin \phi$$

$$w_i = -\bar{u}_i \sin \phi + \bar{w}_i \cos \phi, \quad \text{and} \quad \left(\frac{dw}{ds}\right)_i = -\beta_i$$

Writing these equations in matrix form the transformation matrix for node 1 is

$$\left\{ \begin{array}{c} u_1 \\ w_1 \\ \left(\dfrac{dw}{ds}\right)_1 \end{array} \right\} = \begin{bmatrix} \cos \phi & \sin \phi & 0 \\ -\sin \phi & \cos \phi & 0 \\ 0 & 0 & -1 \end{bmatrix} \left\{ \begin{array}{c} \bar{u}_1 \\ \bar{w}_1 \\ \beta_1 \end{array} \right\} = [T]\{\delta_1\} \qquad (9.9)$$

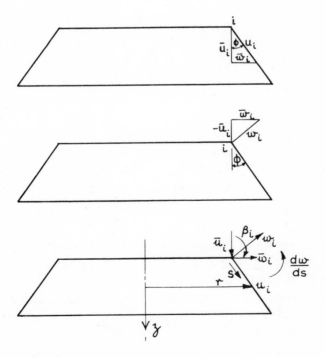

Fig. 9.2. Local to global transformations

For a single element,

$$
\begin{Bmatrix}
u_1 \\
w_1 \\
\left(\dfrac{dw}{ds}\right)_1 \\
\hline
u_2 \\
w_2 \\
\left(\dfrac{dw}{ds}\right)_2
\end{Bmatrix}
=
\left[
\begin{array}{ccc:ccc}
\cos\phi & \sin\phi & 0 & 0 & 0 & 0 \\
-\sin\phi & \cos\phi & 0 & 0 & 0 & 0 \\
0 & 0 & -1 & 0 & 0 & 0 \\
\hdashline
0 & 0 & 0 & \cos\phi & \sin\phi & 0 \\
0 & 0 & 0 & -\sin\phi & \cos\phi & 0 \\
0 & 0 & 0 & 0 & 0 & -1
\end{array}
\right]
\begin{Bmatrix}
\bar{u}_1 \\
\bar{w}_1 \\
\beta_1 \\
\hline
\bar{u}_2 \\
\bar{w}_2 \\
\beta_2
\end{Bmatrix}
$$

$$(9.10)$$

Equation 9.10 can alternatively be written as

$$
\begin{Bmatrix}
u_1 \\
w_1 \\
\left(\dfrac{dw}{ds}\right)_1 \\
\hline
u_2 \\
w_2 \\
\left(\dfrac{dw}{ds}\right)_2
\end{Bmatrix}
=
\left[
\begin{array}{c:c}
[T] & [0] \\
\hdashline
 & \\
\hdashline
[0] & [T]
\end{array}
\right]
\{\delta^e\}
\qquad (9.11)
$$

where $[T]$ is given in equation 9.9 and $\{\delta^e\}$, the element displacement vector, is given in equation 9.1.

It is now possible to write the equation giving the element local displacements $\{{}^u_w\}$ in terms of the global nodal deformation for the element $\{\delta^e\}$. This is done by substituting into equation 9.8 for $\{\alpha\}$ from equation 9.7, and then replacing the local nodal deformations by the global nodal deformations given in equation 9.10 or 9.11. These are written out in full in equation 9.12 and the matrix multiplication is performed to yield finally equation 9.13, which completes step III (see p. 139 for equations 9.12 and 9.13).

$$\begin{Bmatrix} u \\ w \end{Bmatrix} = \begin{bmatrix} 1 & s & 0 & 0 & 0 & 0 \\ 0 & 0 & 1 & s & s^2 & s^3 \end{bmatrix} \begin{bmatrix} 1 & 0 & 0 & 0 & 0 & 0 \\ -\dfrac{1}{L} & 0 & 0 & \dfrac{1}{L} & 0 & 0 \\ 0 & 1 & 0 & 0 & 0 & 0 \\ 0 & 0 & 1 & 0 & 0 & 0 \\ 0 & -\dfrac{3}{L^2} & -\dfrac{2}{L} & 0 & \dfrac{3}{L^2} & -\dfrac{1}{L} \\ 0 & \dfrac{2}{L^3} & \dfrac{1}{L^2} & 0 & -\dfrac{2}{L^3} & \dfrac{1}{L^2} \end{bmatrix} \begin{bmatrix} \cos\phi & \sin\phi & 0 & 0 & 0 & 0 \\ -\sin\phi & \cos\phi & 0 & 0 & 0 & 0 \\ 0 & 0 & -1 & 0 & 0 & 0 \\ 0 & 0 & 0 & \cos\phi & \sin\phi & 0 \\ 0 & 0 & 0 & -\sin\phi & \cos\phi & 0 \\ 0 & 0 & 0 & 0 & 0 & -1 \end{bmatrix} \begin{Bmatrix} \bar u_1 \\ \bar w_1 \\ \beta_1 \\ \bar u_2 \\ \bar w_2 \\ \beta_2 \end{Bmatrix} \quad (9.12)$$

$$= \begin{bmatrix} (1-p)\cos\phi & (1-p)\sin\phi & 0 & p\cos\phi & p\sin\phi & 0 \\ -(1-3p^2+2p^3)\times\sin\phi & (1-3p^2+2p^3)\times\cos\phi & -L(p-2p^2+p^3) & -(3p^2-2p^3)\times\sin\phi & (3p^2-2p^3)\times\cos\phi & -L(-p^2+p^3) \end{bmatrix} \{\delta^e\} \quad (9.13)$$

where $p = s/L$.

Step IV: Relate strains $\{\varepsilon(r, s)\}$ at any point in element to displacements and hence to nodal displacements $\{\delta^e\}$

The components of strain for the middle surface of a conical frustum axisymmetric shell involve extensions and curvatures which are the two in-plane strains ε_s and ε_θ (hoop strain), and the corresponding curvatures χ_s and χ_θ. These are related to the displacements u and w by equation 9.14.

$$\{\varepsilon(r, s)\} = \begin{Bmatrix} \varepsilon_s \\ \\ \varepsilon_\theta \\ \\ \chi_s \\ \\ \chi_\theta \end{Bmatrix} = \begin{bmatrix} \dfrac{d}{ds} & 0 \\ \\ \dfrac{\sin \phi}{r} & \dfrac{\cos \phi}{r} \\ \\ 0 & \dfrac{d^2}{ds^2} \\ \\ 0 & \dfrac{-\sin \phi}{r} \dfrac{d}{ds} \end{bmatrix} \begin{Bmatrix} u \\ w \end{Bmatrix} \tag{9.14}$$

Substituting from equation 9.13 for $\{{u \atop w}\}$ and performing the differentiations indicated in equation 9.14 gives the strain–nodal displacement matrix $[B]$.

$$\{\varepsilon(r, s)\} = [B]\{\delta^e\} \tag{IV}$$

It should be noted that since $p = s/L$,

$$ds = L\,dp \quad \text{and} \quad \frac{d}{ds} = \frac{1}{L}\frac{d}{dp}$$

Matrix $[B]$ is a 4×6 matrix, since it is the product of a 4×2 matrix (equation 9.14) and a 2×6 matrix (equation 9.13), and is given in full in equation 9.15.

Step V: Relate internal stresses $\{\sigma(r, s)\}$ to strains $\{\varepsilon(r, s)\}$ and to nodal displacements $\{\delta^e\}$

In the case of shells it is usual to work in terms of the stress resultants, which are the forces and moments per unit length. For this axisymmetric shell element these resultants consist of N_s, N_θ which

$$[B] =$$

$$
\begin{bmatrix}
\dfrac{-\cos\phi}{L} & \dfrac{-\sin\phi}{L} & 0 & \dfrac{\cos\phi}{L} & \dfrac{\sin\phi}{L} & 0 \\[2ex]
\begin{array}{l}(1-p)(\cos\phi\sin\phi)/r\\ -(1-3p^2+2p^3)\\ \times(\sin 2\phi)/2r\end{array} &
\begin{array}{l}(1-p)\sin^2\phi\\ +(1-3p^2+2p^3)\\ \times(\cos^2\phi)/r\end{array} &
\begin{array}{l}-L(p-2p^2+p^3)\\ \times(\cos\phi)/r\end{array} &
\begin{array}{l}p\sin\phi\cos\phi\\ -(3p^2-2p^3)\\ \times(\sin 2\phi)/2r\end{array} &
\begin{array}{l}p\sin^2\phi\\ +(3p^2-2p^3)\\ \times(\cos^2\phi)/r\end{array} &
\begin{array}{l}-L(-p^2+p^3)\\ \times(\cos\phi)/r\end{array} \\[3ex]
\begin{array}{l}-(1/L^2)(-6+12p)\\ \times\sin\phi\end{array} &
\begin{array}{l}(1/L^2)(-6+12p)\\ \times\cos\phi\end{array} &
-(1/L)(-4+6p) &
\begin{array}{l}-(1/L^2)(6-12p)\\ \times\sin\phi\end{array} &
\begin{array}{l}(1/L^2)(6-12p)\\ \times\cos\phi\end{array} &
-(1/L)(-2+6p) \\[2ex]
\begin{array}{l}(1/L)(-6p+6p^2)\\ \times(\sin^2\phi)/r\end{array} &
\begin{array}{l}-(1/L)(-6p+6p^2)\\ \times(\sin 2\phi)/2r\end{array} &
\begin{array}{l}-(1-4p+3p^2)\\ \times(\sin\phi)/r\end{array} &
\begin{array}{l}(1/L)(6p-6p^2)\\ \times(\sin^2\phi)/r\end{array} &
\begin{array}{l}-(1/L)(6p-6p^2)\\ \times(\sin 2\phi)/2r\end{array} &
\begin{array}{l}-(-2p+3p^2)\\ \times(\sin\phi)/r\end{array}
\end{bmatrix}
$$

$$(9.15)$$

where $p = s/L$.

are the membrane forces per unit length, and M_s, M_θ, the moments per unit length, as shown in Figure 9.1. The stress–strain matrix, $[D]$, is given in equation 9.16.

$$\begin{Bmatrix} N_s \\ N_\theta \\ M_s \\ M_\theta \end{Bmatrix} = \frac{Et}{(1-v^2)} \begin{bmatrix} 1 & v & 0 & 0 \\ v & 1 & 0 & 0 \\ 0 & 0 & \dfrac{t^2}{12} & \dfrac{vt^2}{12} \\ 0 & 0 & \dfrac{vt^2}{12} & \dfrac{t^2}{12} \end{bmatrix} \begin{Bmatrix} \varepsilon_s \\ \varepsilon_\theta \\ \chi_s \\ \chi_\theta \end{Bmatrix} \tag{9.16}$$

or $\{\sigma(r, s)\} = [D]\{\varepsilon(r, s)\}$, where E is Young's modulus of elasticity, v is Poisson's ratio and t is the shell thickness. Substituting for $\{\varepsilon(r, s)\}$ from step IV the stresses in the element can now be related to the nodal displacements,

$$\{\sigma(r, s)\} = [D][B]\{\delta^e\} \tag{V}$$

The product $[D][B]$ of equations 9.15 and 9.16 results in a 4×6 matrix.

Step VI: Replace internal stresses $\{\sigma(r, s)\}$ with statically equivalent nodal forces $\{F^e\}$, relate nodal forces to nodal displacements $\{\delta^e\}$ and hence obtain element stiffness matrix $[K^e]$

The details of this step are the same as those which were developed in Chapter 3 and the final result is given by equation VI.

$$\{F^e\} = [\textstyle\int [B]^T [D][B]\, d(\text{area})]\{\delta^e\} \tag{VI}$$

The only difference between this expression and that given in Chapter 3 is that in this case the integration is taken over the area of the element. The element thickness t is not involved here because the stresses are expressed as stress resultants, forces and moments per unit length.

The element stiffness matrix $[K^e]$ is given by

$$[K^e] = \int_{area} [B]^T [D][B]\, d(\text{area})$$

Referring to Figure 9.1, for this conical frustum $d(\text{area}) = 2\pi r \, ds = 2\pi r L \, dp$, while s varies from 0 to L for the element, p varies from 0 to 1 since $p = s/L$. Hence the element stiffness matrix is

$$[K^e] = \int_0^1 [B]^T [D][B] 2\pi r L \, dp \qquad (9.17)$$

In this case r must be expressed in terms of s and hence p. Since the terms in the $[B]$ matrix depend upon r and p, the matrix multiplications must be performed and the integration carried out term by term. This is a rather lengthy task and has not been done here for the general case of any angle ϕ.[1] However, later in this chapter, the results are presented for two particular cases that are frequently encountered in engineering, a circular cylindrical shell and a circular flat plate.

Step VII: Establish stress–displacement matrix $[H]$

The final step is the determination of the stresses at any point in the element from the element nodal displacements. The relationship given in equation V enables this to be done.

$$\{\sigma(r, s)\} = [D][B]\{\delta^e\} \qquad (V)$$

or

$$\{\sigma(r, s)\} = [H]\{\delta^e\} \qquad (VII)$$

The stress–displacement matrix $[H]$ is obtained by pre-multiplying the matrix $[B]$ given in equation 9.15 by the $[D]$ matrix given in equation 9.16. The solution for the general case of any angle ϕ is given in reference 1.

Examples

a: Circular cylindrical shell: $\phi = 0$

Figure 9.3 shows a typical element taken from a circular cylindrical shell where r is now a constant. The required element stiffness matrix $[K^e]$ (step VI) and its stress–displacement matrix $[H]$ (step VII) can be obtained in one of two ways, either by substituting $\phi = 0$ in the matrix $[B]$ in equation 9.15 and then performing the operations indicated in steps VI and VII, or alternatively by returning to step 3, putting $\phi=0$ in the transformation matrix of equations 9.9 or 9.10 and in the strain displacement matrix of equation 9.14 and then proceeding through the succeeding steps. Employing both methods provides a valuable check on the answers.

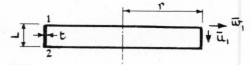

Fig. 9.3. Cylindrical shell element

Here $\phi=0$ is substituted in matrix $[B]$ of equation 9.15, which is written below in equation 9.18 for the cylindrical shell element.

$$[B] =$$

$$
\begin{bmatrix}
-\dfrac{1}{L} & 0 & 0 & \dfrac{1}{L} & 0 & 0 \\[2ex]
0 & \dfrac{(1-3p^2+2p^3)}{r} & -\dfrac{L}{r}(p-2p^2+p^3) & 0 & \dfrac{(3p^2-2p^3)}{r} & -\dfrac{L}{r}(-p^2+p^3) \\[2ex]
0 & \dfrac{(-6+12p)}{L^2} & -\dfrac{1}{L}(-4+6p) & 0 & \dfrac{1}{L^2}(6-12p) & -\dfrac{1}{L}(-2+6p) \\[2ex]
0 & 0 & 0 & 0 & 0 & 0
\end{bmatrix}
$$

$$(9.18)$$

where $p=s/L$. It should be noted from equation 9.16 that the $[D]$ matrix is independent of ϕ.

$$[D][B] = \frac{Et}{(1-v^2)} \times$$

$$
\begin{bmatrix}
-\dfrac{1}{L} & \dfrac{v}{r}(1-3p^2+2p^3) & -\dfrac{vL(p-2p^2+p^3)}{r} & \dfrac{1}{L} & \dfrac{v}{r}(3p^2-2p^3) & -\dfrac{vL}{r}(-p^2+p^3) \\[2ex]
-\dfrac{v}{L} & \dfrac{1}{r}(1-3p^2+2p^3) & -\dfrac{L}{r}(p-2p^2+p^3) & \dfrac{v}{L} & \dfrac{1}{r}(3p^2-2p^3) & -\dfrac{L}{r}(-p^2+p^3) \\[2ex]
0 & \dfrac{t^2(-6+12p)}{12L^2} & -\dfrac{t^2(-4+6p)}{12L} & 0 & \dfrac{t^2(6-12p)}{12L^2} & -\dfrac{t^2(-2+6p)}{12L} \\[2ex]
0 & \dfrac{vt^2(-6+12p)}{12L^2} & -\dfrac{vt^2(-4+6p)}{12L} & 0 & \dfrac{vt^2(6-12p)}{12L^2} & -\dfrac{vt^2(-2+6p)}{12L}
\end{bmatrix}
$$

$$(9.19)$$

Since matrices $[B]$ and $[D]$ are now known the element stiffness matrix $[K^e]$ can be readily obtained using equation VI. First the product $[D][B]$ is formed (since this is equal to the stress matrix $[H]$ required in step VII), and is given in equation 9.19.

The next operation is to pre-multiply this equation by $[B]^T$, the result being given in equation 9.20.

$$[B^T][D][B] = \frac{Et}{(1-v^2)} \begin{bmatrix} b_{11} & b_{12} & b_{13} & b_{14} & b_{15} & b_{16} \\ b_{21} & b_{22} & b_{23} & b_{24} & b_{25} & b_{26} \\ b_{31} & b_{32} & b_{33} & b_{34} & b_{35} & b_{36} \\ b_{41} & b_{42} & b_{43} & b_{44} & b_{45} & b_{46} \\ b_{51} & b_{52} & b_{53} & b_{54} & b_{55} & b_{56} \\ b_{61} & b_{62} & b_{63} & b_{64} & b_{65} & b_{66} \end{bmatrix} \qquad (9.20)$$

$$b_{11} = \frac{1}{L^2} \quad b_{12} = -\frac{v}{rL}(1-3p^2+2p^3) \quad b_{13} = \frac{v}{r}(p-2p^2+p^3)$$

$$b_{14} = -\frac{1}{L^2} \quad b_{15} = -\frac{v}{rL}(3p^2-2p^3) \quad b_{16} = \frac{v}{r}(-p^2+p^3)$$

$$b_{22} = \frac{1}{r^2}(1-3p^2+2p^3)^2 + \frac{t^2}{12L^4}(-6+12p)^2$$

$$b_{23} = -\frac{L}{r^2}(1-3p^2+2p^3)(p-2p^2+p^3) - \frac{t^2}{12L^3}(-6+12p)(-4+6p)$$

$$b_{24} = \frac{v}{rL}(1-3p^2+2p^3)$$

$$b_{25} = \frac{1}{r^2}(1-3p^2+2p^3)(3p^2-2p^3) + \frac{t^2}{12L^4}(-6+12p)(6-12p)$$

$$b_{26} = -\frac{L}{r^2}(1-3p^2+2p^3)(-p^2+p^3) - \frac{t^2}{12L^3}(-6+12p)(-2+6p)$$

$$b_{33} = \frac{L^2}{r^2}(p-2p^2+p^3)^2 + \frac{t^2}{12L^2}(-4+6p)^2$$

$$b_{34} = -\frac{v}{r}(p-2p^2+p^3)$$

$$b_{35} = -\frac{L}{r^2}(p-2p^2+p^3)(3p^2-2p^3)-\frac{t^2}{12L^3}(-4+6p)(6-12p)$$

$$b_{36} = \frac{L^2}{r^2}(p-2p^2+p^3)(-p^2+p^3)+\frac{t^2}{12L^2}(-4+6p)(-2+6p)$$

$$b_{44} = \frac{1}{L^2} \quad b_{45} = \frac{v}{rL}(3p^2-2p^3) \quad b_{46} = -\frac{v}{r}(-p^2+p^3)$$

$$b_{55} = \frac{1}{r^2}(3p^2-2p^3)^2+\frac{t^2}{12L^4}(6-12p)^2$$

$$b_{56} = -\frac{L}{r^2}(3p^2-2p^3)(-p^2+p^3)-\frac{t^2}{12L^3}(6-12p)(-2+6p)$$

$$b_{66} = \frac{L^2}{r^2}(-p^2+p^3)^2+\frac{t^2}{12L^2}(-2+6p)^2$$

The matrix in equation 9.20 is symmetric and so only the terms on and above the leading diagonal are given.

The element stiffness matrix $[K^e]$ is given in equation 9.17 as

$$[K^e] = \int_0^1 [B]^T[D][B]2\pi rL\,dp$$

The value of the product of $[B]^T[D][B]$ has already been determined in equation 9.20. Thus to obtain the terms in $[K^e]$ each term in equation 9.20 must be multiplied by $2\pi rL$ and integrated with respect to p between the limits 0 and 1. Writing $[K^e]$ as in equation 9.21,

$$[K^e] = 2\pi\left(\frac{Et}{1-v^2}\right)\begin{bmatrix} k_{11} & k_{12} & k_{13} & k_{14} & k_{15} & k_{16} \\ k_{21} & k_{22} & k_{23} & k_{24} & k_{25} & k_{26} \\ k_{31} & k_{32} & k_{33} & k_{34} & k_{35} & k_{36} \\ k_{41} & k_{42} & k_{43} & k_{44} & k_{45} & k_{46} \\ k_{51} & k_{52} & k_{53} & k_{54} & k_{55} & k_{56} \\ k_{61} & k_{62} & k_{63} & k_{64} & k_{65} & k_{66} \end{bmatrix} \qquad (9.21)$$

From equation 9.17, each term in equation 9.21 may be found using equation 9.20 as follows.

$$k_{11} = \int_0^1 b_{11}\, rL\, dp = \int_0^1 \frac{1}{L^2}\, rL\, dp = \frac{r}{L};$$

$$k_{12} = \int_0^1 b_{12}\, rL\, dp = \int_0^1 -\frac{v}{rL}(1-3p^2+2p^3)rL\, dp = -\frac{v}{2};\ \text{etc.}$$

All the values for the k_{ij} terms obtained in this way are listed below. Note that $2\pi[Et/(1-v^2)]$ is a constant multiplier for all these terms, as indicated in equation 9.21.

$$k_{11} = \frac{r}{L} \quad k_{12} = -\frac{v}{2} \quad k_{13} = \frac{vL}{12} \quad k_{14} = -\frac{r}{L} = -k_{11}$$

$$k_{15} = -\frac{v}{2} = k_{12} \quad k_{16} = -\frac{vL}{12} = -k_{13}$$

$$k_{22} = \frac{13}{35}\frac{L}{r}+\frac{rt^2}{L^3} \quad k_{23} = -\frac{11}{210}\frac{L^2}{r}-\frac{rt^2}{2L^2} \quad k_{24} = \frac{v}{2} = -k_{12}$$

$$k_{25} = \frac{9}{70}\frac{L}{r}-\frac{rt^2}{L^3} \quad k_{26} = \frac{13}{420}\frac{L^2}{r}-\frac{1}{2}\frac{rt^2}{L^2}$$

$$k_{33} = \frac{1}{105}\frac{L^3}{r}+\frac{1}{3}\frac{rt^2}{L} \quad k_{34} = -\frac{vL}{12} = -k_{13}$$

$$k_{35} = -\frac{13}{420}\frac{L^2}{r}+\frac{1}{2}\frac{rt^2}{L^2} = -k_{26} \quad k_{36} = -\frac{1}{140}\frac{L^3}{r}+\frac{1}{6}\frac{rt^2}{L}$$

$$k_{44} = \frac{r}{L} = k_{11} \quad k_{45} = \frac{v}{2} = -k_{12} \quad k_{46} = \frac{vL}{12} = k_{13}$$

$$k_{55} = \frac{13}{35}\frac{L}{r}+\frac{rt^2}{L^3} = k_{22} \quad k_{56} = \frac{11}{210}\frac{L^2}{r}+\frac{rt^2}{2L^2} = -k_{23}$$

$$k_{66} = \frac{1}{105}\frac{L^3}{r}+\frac{1}{3}\frac{rt^2}{L} = k_{33}$$

Again this matrix is symmetric and only the leading diagonal and terms above it are presented, since $k_{ij}=k_{ji}$.

Finally, the stress–displacement matrix $[H]$ is required. This has already been evaluated and appears in equation 9.19, since $[H] =[D][B]$. Thus when the nodal displacements have been obtained, the stresses at any point in the element can be determined. The point considered is specified by the value of $p(=s/L)$, which equals zero at node 1 and unity at node 2, varying linearly in between. For example to determine the stresses at node 1, by substituting $p=0$ in equation 9.19, equation 9.22 is obtained for the stresses at node 1 in terms of the element nodal displacements.

$$
\begin{Bmatrix} N_{s1} \\ N_{\theta 1} \\ M_{s1} \\ M_{\theta 1} \end{Bmatrix} = \left(\frac{Et}{1-v^2} \right) \begin{bmatrix} -\dfrac{1}{L} & \dfrac{v}{r} & 0 & \dfrac{1}{L} & 0 & 0 \\ -\dfrac{v}{L} & \dfrac{1}{r} & 0 & \dfrac{v}{L} & 0 & 0 \\ 0 & -\dfrac{t^2}{2L^2} & \dfrac{t^2}{3L} & 0 & \dfrac{t^2}{2L^2} & \dfrac{t^2}{6L} \\ 0 & -\dfrac{vt^2}{2L^2} & \dfrac{vt^2}{3L} & 0 & \dfrac{vt^2}{2L^2} & \dfrac{vt^2}{6L} \end{bmatrix} \begin{Bmatrix} \bar{u}_1 \\ \bar{w}_1 \\ \beta_1 \\ \bar{u}_2 \\ \bar{w}_2 \\ \beta_2 \end{Bmatrix}
$$

(9.22)

At node 2, $p=1$, and hence the stresses are given by equation 9.23.

$$
\begin{Bmatrix} N_{s2} \\ N_{\theta 2} \\ M_{s2} \\ M_{\theta 2} \end{Bmatrix} = \left(\frac{Et}{1-v^2} \right) \begin{bmatrix} -\dfrac{1}{L} & 0 & 0 & \dfrac{1}{L} & \dfrac{v}{r} & 0 \\ -\dfrac{v}{L} & 0 & 0 & \dfrac{v}{L} & \dfrac{1}{r} & 0 \\ 0 & +\dfrac{t^2}{2L^2} & -\dfrac{t^2}{6L} & 0 & -\dfrac{t^2}{2L^2} & -\dfrac{t^2}{3L} \\ 0 & +\dfrac{vt^2}{2L^2} & \dfrac{vt^2}{6L} & 0 & -\dfrac{vt^2}{2L^2} & -\dfrac{vt^2}{3L} \end{bmatrix} \begin{Bmatrix} \bar{u}_1 \\ \bar{w}_1 \\ \beta_1 \\ \bar{u}_2 \\ \bar{w}_2 \\ \beta_2 \end{Bmatrix}
$$

(9.23)

Similarly, the stresses at any other point are readily determined by simply substituting the appropriate value of p for that point.

b: Circular flat plate: $\phi = 90°$

Figure 9.4 shows a typical element taken from a flat plate. Here the radius r varies according to $r = r_1 + Lp$, where r_1 is the radius at node 1 and L is the distance between the nodes. Substituting $\phi = 90°$ in equation 9.15 gives equation 9.24.

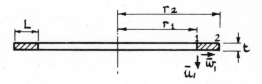

Fig. 9.4. Circular plate element

$$= \begin{bmatrix} 0 & -\dfrac{1}{L} & 0 & 0 & \dfrac{1}{L} & 0 \\[2ex] 0 & \dfrac{1-p}{r} & 0 & 0 & \dfrac{p}{r} & 0 \\[2ex] -\dfrac{1}{L^2}(-6+12p) & 0 & -\dfrac{1}{L}(-4+6p) & -\dfrac{1}{L^2}(6-12p) & 0 & -\dfrac{1}{L}(-2+6p) \\[2ex] \dfrac{1}{rL}(-6p+6p^2) & 0 & \dfrac{1}{r}(1-4p+3p^2) & \dfrac{1}{rL}(6p-6p^2) & 0 & \dfrac{1}{r}(-2p+3p^2) \end{bmatrix}$$

$$(9.24)$$

Next the product $[D][B]$, as given in equation 9.25 (p. 150), is formed.

The product $[B]^T[D][B]$ is presented in the same form as equation 9.20, where now

$$b_{11} = \frac{t^2}{12L^4}(36 - 144p + 144p^2) - \frac{vt^2}{6rL^3}(36p - 108p^2 + 72p^3)$$

$$+ \frac{t^2}{12r^2L^2}(36p^2 - 72p^3 + 36p^4)$$

$$b_{12} = 0$$

$$
[D][B] = \frac{Et}{(1-v^2)}
\begin{bmatrix}
-\dfrac{1}{L}+\dfrac{v(1-p)}{r} & 0 & 0 & \dfrac{1}{L}+\dfrac{vp}{r} & 0 & 0 \\[2ex]
-\dfrac{v}{L}+\dfrac{(1-p)}{r} & 0 & 0 & \dfrac{v}{L}+\dfrac{p}{r} & 0 & 0 \\[2ex]
0 & \left(-\dfrac{t^2}{12L^2}(-6+12p)+\dfrac{vt^2}{12rL}(-6p+6p^2)\right) & \left(-\dfrac{t^2}{12L}(-4+6p)+\dfrac{vt^2}{12r}(1-4p+3p^2)\right) & 0 & \left(-\dfrac{t^2}{12L^2}(6-12p)+\dfrac{vt^2}{12rL}(6p-6p^2)\right) & \left(-\dfrac{t^2}{12L}(-2+6p)+\dfrac{vt^2}{12r}(-2p+3p^2)\right) \\[3ex]
0 & \left(-\dfrac{vt^2}{12L^2}(-6+12p)+\dfrac{t^2}{12rL}(-6p+6p^2)\right) & \left(-\dfrac{vt^2}{12L}(-4+6p)+\dfrac{t^2}{12r}(1-4p+3p^2)\right) & 0 & \left(-\dfrac{vt^2}{12L^2}(6-12p)+\dfrac{t^2}{12rL}(6p-6p^2)\right) & \left(-\dfrac{vt^2}{12L}(-2+6p)+\dfrac{t^2}{12r}(-2p+3p^2)\right)
\end{bmatrix}
\tag{9.25}
$$

$$b_{13} = \frac{t^2}{12L^3}(24 - 84p + 72p^2) - \frac{vt^2}{12rL^2}(-6 + 60p - 126p^2 + 72p^3)$$

$$+ \frac{t^2}{12r^2L}(-6p + 30p^2 - 42p^3 + 18p^4)$$

$$b_{14} = -b_{11} \quad b_{15} = 0$$

$$b_{16} = -\frac{t^2}{12L^3}(12 - 60p + 72p^2) - \frac{vt^2}{12rL^2}(24p - 90p^2 + 72p^3)$$

$$+ \frac{t^2}{12r^2L}(12p^2 - 30p^3 + 18p^4)$$

$$b_{22} = \frac{1}{L^2} - \frac{2v(1-p)}{rL} + \frac{1 - 2p + p^2}{r^2} \quad b_{23} = 0 \quad b_{24} = 0$$

$$b_{25} = -\frac{1}{L^2} + \frac{v}{rL}(1 - 2p) + \frac{p(1-p)}{r^2} \quad b_{26} = 0$$

$$b_{33} = \frac{t^2}{12L^2}(16 - 48p + 36p^2) - \frac{vt^2}{6rL}(-4 + 22p - 36p^2 + 18p^3)$$

$$+ \frac{t^2}{12r^2}(1 - 8p + 22p^2 - 24p^3 + 9p^4)$$

$$b_{34} = -b_{13} \quad b_{35} = 0$$

$$b_{36} = \frac{t^2}{12L^2}(8 - 36p + 36p^2) - \frac{vt^2}{12rL}(-2 + 22p - 54p^2 + 36p^3)$$

$$+ \frac{t^2}{12r^2}(-2p + 11p^2 - 18p^3 + 9p^4)$$

$$b_{44} = b_{11} \quad b_{45} = 0 \quad b_{46} = -b_{16}$$

$$b_{55} = \frac{1}{L^2} + \frac{2vp}{rL} + \frac{p^2}{r^2} \quad b_{56} = 0$$

$$b_{66} = \frac{t^2}{12L^2}(4 - 24p + 36p^2) - \frac{vt^2}{6rL}(4p - 18p^2 + 18p^3)$$

$$+ \frac{t^2}{12r^2}(4p^2 - 12p^3 + 9p^4)$$

The terms of the element stiffness matrix $[K^e]$ are presented in the same form as equation 9.21, where

$$k_{11} = \int_0^1 b_{11} rL \, dp$$

$$= \int_0^1 b_{11}(r_1 + Lp) L \, dp$$

since

$$r = r_1 + s = r_1 + Lp$$

Substituting for b_{11} from above and performing the integration term by term, k_{11} is obtained, the remaining terms being determined similarly. The results of the integration are given below. Since the element stiffness matrix is symmetric only the terms on and above the leading diagonal are presented.

$$k_{11} = \frac{t^2}{12L^2} \left[6 - 60\frac{r_1}{L} - 216\left(\frac{r_1}{L}\right)^2 - 144\left(\frac{r_1}{L}\right)^3 \right.$$
$$+ \left\{ 18\left(\frac{r_1}{L}\right)^2 + 108\left(\frac{r_1}{L}\right)^3 + 108\left(\frac{r_1}{L}\right)^4 \right\} \left\{ \frac{r_2^2 - r_1^2}{r_1^2} \right\}$$
$$- \left\{ 24\left(\frac{r_1}{L}\right)^3 + 48\left(\frac{r_1}{L}\right)^4 \right\} \left\{ \frac{r_2^3 - r_1^3}{r_1^3} \right\} + 9\left(\frac{r_1}{L}\right)^4$$
$$\times \left(\frac{r_2^4 - r_1^4}{r_1^4} \right)$$
$$\left. + \left\{ 36\left(\frac{r_1}{L}\right)^2 + 72\left(\frac{r_1}{L}\right)^3 + 36\left(\frac{r_1}{L}\right)^4 \right\} \ln\left(\frac{r_2}{r_1}\right) \right]$$

$$k_{12} = 0$$

$$k_{13} = \frac{t^2}{12L} \left[4 + 54\left(\frac{r_1}{L}\right) + 126\left(\frac{r_1}{L}\right)^2 + 72\left(\frac{r_1}{L}\right)^3 \right.$$
$$- \left\{ 15\left(\frac{r_1}{L}\right)^2 + 63\left(\frac{r_1}{L}\right)^3 + 54\left(\frac{r_1}{L}\right)^4 \right\} \left\{ \frac{r_2^2 - r_1^2}{r_1^2} \right\}$$
$$+ \left\{ 14\left(\frac{r_1}{L}\right)^3 + 24\left(\frac{r_1}{L}\right)^4 \right\} \left\{ \frac{r_2^3 - r_1^3}{r_1^3} \right\} - 4 \cdot 5\left(\frac{r_1}{L}\right)^4$$
$$\times \left(\frac{r_2^4 - r_1^4}{r_1^4} \right)$$
$$\left. - \left\{ 6\left(\frac{r_1}{L}\right) + 30\left(\frac{r_1}{L}\right)^2 + 42\left(\frac{r_1}{L}\right)^3 + 18\left(\frac{r_1}{L}\right)^4 \right\} \ln\frac{r_2}{r_1} \right]$$

$$k_{14} = -k_{11} \quad k_{15} = 0$$

$$
k_{16} = \frac{t^2}{12L}\left[4 - 18\left(\frac{r_1}{L}\right) - 90\left(\frac{r_1}{L}\right)^2 - 72\left(\frac{r_1}{L}\right)^3\right.
$$
$$
+ \left\{6\left(\frac{r_1}{L}\right)^2 + 45\left(\frac{r_1}{L}\right)^3 + 54\left(\frac{r_1}{L}\right)^4\right\}\left\{\frac{r_2{}^2 - r_1{}^2}{r_1{}^2}\right\}
$$
$$
- \left\{10\left(\frac{r_1}{L}\right)^3 + 24\left(\frac{r_1}{L}\right)^4\right\}\left\{\frac{r_2{}^3 - r_1{}^3}{r_1{}^3}\right\} + 4\cdot5\left(\frac{r_1}{L}\right)^4
$$
$$
\times \left(\frac{r_2{}^4 - r_1{}^4}{r_1{}^4}\right)
$$
$$
\left. + \left\{12\left(\frac{r_1}{L}\right)^2 + 30\left(\frac{r_1}{L}\right)^3 + 18\left(\frac{r_1}{L}\right)^4\right\}\ln\left(\frac{r_2}{r_1}\right)\right]
$$

$$
k_{22} = -1\cdot5 - v - \frac{r_1}{L} + 0\cdot5\left(\frac{r_1}{L}\right)^2\left(\frac{r_2{}^2 - r_1{}^2}{r_1{}^2}\right)
$$
$$
+ \left\{1 + 2\left(\frac{r_1}{L}\right) + \left(\frac{r_1}{L}\right)^2\right\}\ln\left(\frac{r_2}{r_1}\right)
$$

$$k_{23} = 0 \quad k_{24} = 0$$

$$
k_{25} = 0\cdot5 + \frac{r_1}{L} - 0\cdot5\left(\frac{r_1}{L}\right)^2\left(\frac{r_2{}^2 - r_1{}^2}{r_1{}^2}\right) - \left\{\left(\frac{r_1}{L}\right) + \left(\frac{r_1}{L}\right)^2\right\}\ln\left(\frac{r_2}{r_1}\right)
$$

$$k_{26} = 0$$

$$
k_{33} = \frac{t^2}{12}\left[-7 - v - 40\left(\frac{r_1}{L}\right) - 72\left(\frac{r_1}{L}\right)^2 - 36\left(\frac{r_1}{L}\right)^3\right.
$$
$$
+ \left\{11\left(\frac{r_1}{L}\right)^2 + 36\left(\frac{r_1}{L}\right)^3 + 27\left(\frac{r_1}{L}\right)^4\right\}\left\{\frac{r_2{}^2 - r_1{}^2}{r_1{}^2}\right\}
$$
$$
- \left\{8\left(\frac{r_1}{L}\right)^3 + 12\left(\frac{r_1}{L}\right)^4\right\}\left\{\frac{r_2{}^3 - r_1{}^3}{r_1{}^3}\right\} + 2\cdot25\left(\frac{r_1}{L}\right)^4
$$
$$
\times \left(\frac{r_2{}^4 - r_1{}^4}{r_1{}^4}\right)
$$
$$
+ \left\{1 + 8\left(\frac{r_1}{L}\right) + 22\left(\frac{r_1}{L}\right)^2 + 24\left(\frac{r_1}{L}\right)^3 + 9\left(\frac{r_1}{L}\right)^4\right\}
$$
$$
\left. \times \ln\left(\frac{r_2}{r_1}\right)\right]
$$

$$k_{34} = -k_{13} \quad k_{35} = 0$$

$$k_{36} = \frac{t^2}{12} \left[-1 - 20 \left(\frac{r_1}{L} \right) - 54 \left(\frac{r_1}{L} \right)^2 - 36 \left(\frac{r_1}{L} \right)^3 \right.$$

$$+ \left\{ 5 \cdot 5 \left(\frac{r_1}{L} \right)^2 + 27 \left(\frac{r_1}{L} \right)^3 + 27 \left(\frac{r_1}{L} \right)^4 \right\} \left\{ \frac{r_2^2 - r_1^2}{r_1^2} \right\}$$

$$- \left\{ 6 \left(\frac{r_1}{L} \right)^3 + 12 \left(\frac{r_1}{L} \right)^4 \right\} \left\{ \frac{r_2^3 - r_1^3}{r_1^3} \right\} + 2 \cdot 25 \left(\frac{r_1}{L} \right)^4$$

$$\times \left(\frac{r_2^4 - r_1^4}{r_1^4} \right)$$

$$\left. + \left\{ 2 \left(\frac{r_1}{L} \right) + 11 \left(\frac{r_1}{L} \right)^2 + 18 \left(\frac{r_1}{L} \right)^3 + 9 \left(\frac{r_1}{L} \right)^4 \right\} \ln \left(\frac{r_2}{r_1} \right) \right]$$

$$k_{44} = k_{11} \quad k_{45} = 0 \quad k_{46} = -k_{16}$$

$$k_{55} = 0 \cdot 5 + v - \left(\frac{r_1}{L} \right) + 0 \cdot 5 \left(\frac{r_1}{L} \right)^2 \left(\frac{r_2^2 - r_1^2}{r_1^2} \right) + \left(\frac{r_1}{L} \right)^2 \ln \left(\frac{r_2}{r_1} \right) \right]$$

$$k_{56} = 0$$

$$k_{66} = \frac{t^2}{12} \left[3 + v - 4 \left(\frac{r_1}{L} \right) - 36 \left(\frac{r_1}{L} \right)^2 - 36 \left(\frac{r_1}{L} \right)^3 \right.$$

$$+ \left\{ 2 \left(\frac{r_1}{L} \right)^2 + 18 \left(\frac{r_1}{L} \right)^3 + 27 \left(\frac{r_1}{L} \right)^4 \right\} \left\{ \frac{r_2^2 - r_1^2}{r_1^2} \right\}$$

$$- \left\{ 4 \left(\frac{r_1}{L} \right)^3 + 12 \left(\frac{r_1}{L} \right)^4 \right\} \left\{ \frac{r_2^3 - r_1^3}{r_1^3} \right\} + 2 \cdot 25 \left(\frac{r_1}{L} \right)^4$$

$$\times \left(\frac{r_2^4 - r_1^4}{r_1^4} \right)$$

$$\left. + \left\{ 4 \left(\frac{r_1}{L} \right)^2 + 12 \left(\frac{r_1}{L} \right)^3 + 9 \left(\frac{r_1}{L} \right)^4 \right\} \ln \left(\frac{r_2}{r_1} \right) \right]$$

This completes the circular flat plate element stiffness matrix $[K^e]$.

Finally the stress–displacement matrix $[H]$ is required. This has already been obtained and is given in equation 9.25, since $[H]$ is the product of $[D]$ and $[B]$.

The stresses at any point are determined by substituting the value of p for that point into equation 9.25 and multiplying by the element nodal displacements. In equations 9.26 and 9.27 the values at node 1 where p is zero and at node 2 where p is unity are given.

$$
\begin{Bmatrix} N_{s1} \\ N_{\theta 1} \\ M_{s1} \\ M_{\theta 1} \end{Bmatrix} = \left(\frac{Et}{1-v^2}\right)
\begin{bmatrix}
0 & -\dfrac{1}{L}+\dfrac{v}{r_1} & 0 & 0 & \dfrac{1}{L} & 0 \\[2mm]
0 & -\dfrac{v}{L}+\dfrac{1}{r_1} & 0 & 0 & \dfrac{v}{L} & 0 \\[2mm]
\dfrac{t^2}{2L^2} & 0 & \dfrac{t}{3L}+\dfrac{vt^2}{12r_1} & -\dfrac{t^2}{2L^2} & 0 & \dfrac{t^2}{6L} \\[2mm]
\dfrac{vt^2}{2L^2} & 0 & \dfrac{vt^2}{3L}+\dfrac{t^2}{12r_1} & -\dfrac{vt^2}{2L^2} & 0 & \dfrac{vt^2}{6L}
\end{bmatrix}
\begin{Bmatrix} \bar{u}_1 \\ \bar{w}_1 \\ \beta_1 \\ \bar{u}_2 \\ \bar{w}_2 \\ \beta_2 \end{Bmatrix} \quad (9.26)
$$

$$
\begin{Bmatrix} N_{s2} \\ N_{\theta 2} \\ M_{s2} \\ M_{\theta 2} \end{Bmatrix} = \left(\frac{Et}{1-v^2}\right)
\begin{bmatrix}
0 & -\dfrac{1}{L} & 0 & 0 & \dfrac{1}{L}+\dfrac{v}{r_2} & 0 \\[2mm]
0 & -\dfrac{v}{L} & 0 & 0 & \dfrac{v}{L}+\dfrac{1}{r_2} & 0 \\[2mm]
-\dfrac{t^2}{2L^2} & 0 & -\dfrac{t^2}{6L} & \dfrac{t^2}{2L^2} & 0 & -\dfrac{t}{3L}+\dfrac{vt^2}{12r_2} \\[2mm]
-\dfrac{vt^2}{2L^2} & 0 & -\dfrac{vt^2}{6L} & \dfrac{vt^2}{2L^2} & 0 & -\dfrac{vt^2}{3L}+\dfrac{t^2}{12r_2}
\end{bmatrix}
\begin{Bmatrix} \bar{u}_1 \\ \bar{w}_1 \\ \beta_1 \\ \bar{u}_2 \\ \bar{w}_2 \\ \beta_2 \end{Bmatrix} \quad (9.27)
$$

Applications of axisymmetric shell element

a: Circular cylindrical shell

This element is applied to a cylindrical shell fixed at one end and loaded at the free end. A solution to this problem has been obtained by the authors using the classical theory.[3] The loading case considered is that of unit radial end loading at the free end while the other end is built in.

Figure 9.5 shows the cylinder and the manner in which it has been idealised for three different finite element sub-divisions. It should be noted that the mesh is graded so that there are more elements near the loaded point since it is in this region that the stresses and deflections change most rapidly.

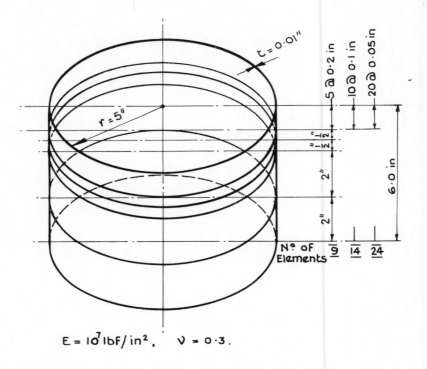

$$E = 10^7 \, lbf/in^2, \quad \nu = 0.3.$$

Fig. 9.5. Idealisation of cylinder using axisymmetric shell elements

Figure 9.6 shows how the radial deflection varies with axial position for the classical theory[3] (shown by the continuous line). Also shown are the finite element results for the 24-element idealisation. The table inset in the figure shows the values of maximum deflection obtained for each idealisation, together with the percentage differences from the theoretical results.[3] It is seen that the 9-element case gives an answer only 2% too low while the 24-element case yields an answer within 0.2%. Figure 9.7 shows the variation in the meridional moment along the tube. Here the errors are greater than for the displacements, being as large as 12% too high in the region of maximum moment for the 9-element case, but falling to within 2% for the 24-element idealisation. Thus it is seen yet again that care must always be taken to ensure that a sufficiently fine mesh is employed but that, if this is done, results to a high degree of accuracy may be obtained.

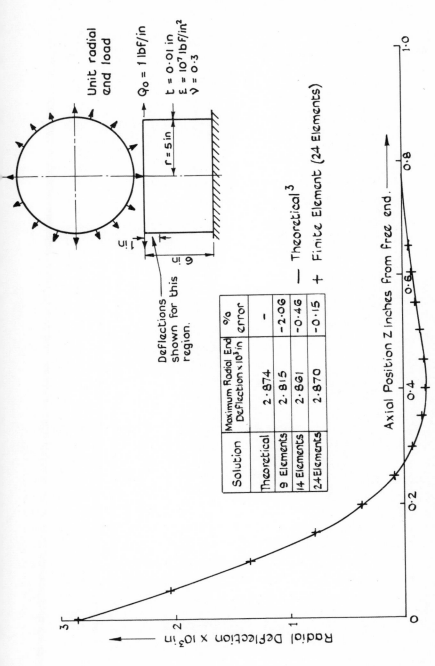

Fig. 9.6. Variation in radial deflection in tube having radial end load

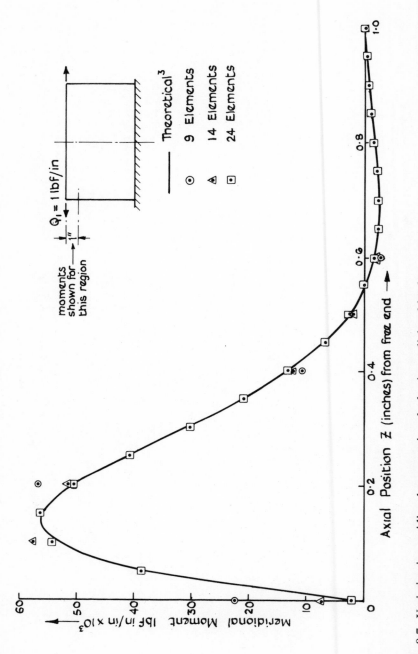

Fig. 9.7. Variation in meridional moment in tube having radial end load

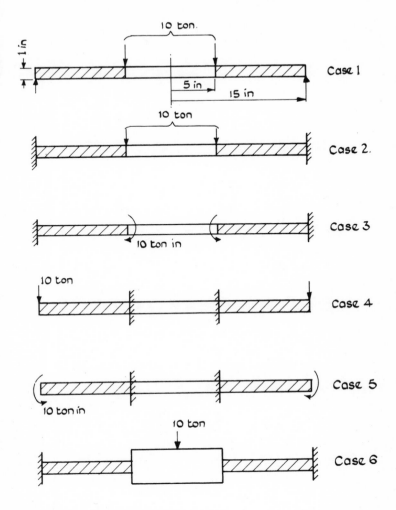

Fig. 9.8. Loading cases for flat circular plate

b: Circular flat plate

This element is used for analysing a circular flat plate under the various loading and support conditions shown in Figure 9.8. The finite element idealisations employed are shown in Figure 9.9, that for the 20-element case including graded sub-divisions as well as

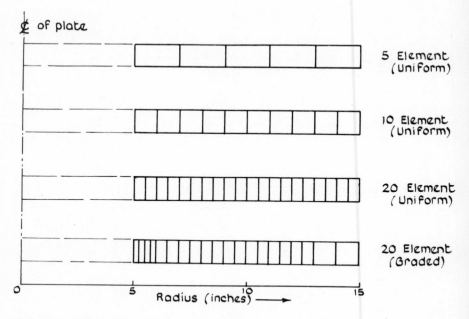

Fig. 9.9. Half section of circular plate idealisations employed

Loading case	Maximum deflection (in)				
	Theoretical	5–element	10–element	20–element	20–element (graded mesh)
1	0·1232	0·1179 −4·3%	0·1217 −1·2%	0·1227 −0·4%	0·1223 −0·7%
2	0·02888	0·02833 −1·9%	0·02874 −0·5%	0·02885 −0·1%	0·02883 −0·2%
3	0·005020	0·004712 −6·1%	0·004936 −1·7%	0·005001 −0·4%	0·005002 −0·4%
4	0·03509	0·03324 −5·3%	0·03466 −1·3%	0·03505 −0·1%	0·03510 0%
5	0·00455	0·004425 −2·7%	0·004519 −0·7%	0·004544 −0·14%	0·004547 −0·07%
6	0·01040	0·009604 −7·7%	0·01020 −2%	0·01036 −0·4%	0·01038 −0·2%

Table 9.1. Circular flat plate: Maximum deflections and percentage errors.

uniform sub-divisions. Tables 9.1 and 9.2 compare the results for the maximum deflections and stresses as obtained from the finite element solutions with those obtained using plate bending theory.[3,4] The variation of the accuracy of the solution with the number of elements, together with the effect of grading the idealisation, is well demonstrated. For loading cases 1, 2 and 3 the graded mesh leads to a lower accuracy in the maximum deflection (Table 9.1) than the uniform mesh while in cases 4, 5 and 6 the accuracy is improved. This is because of the support conditions: the refined mesh in the region of greater fixity gives improved results in cases 4, 5 and 6. For the stresses, Table 9.2 shows that the refined mesh always gives higher accuracy at the inner radius where the stresses are highest, but that at the outer radius (loading case 6) the graded mesh gives a similar accuracy to that of the 10-element sub-division in which the size of the outer element is the same, as can be seen in Figure 9.9. Hence, once again it is clear that care must be taken to ensure that a suffi-

| Loading | Maximum stresses (ton/in^2) | | | | | Stress |
	Theoretical	5-element	10-element	20-element	20-element graded mesh	
1	18·8	14·99 −20·2%	16·72 −11%	17·65 −6·1%	18·02 −4·1%	σ_θ inner
2	6·73	5.12 −23·8%	5·861 −12·7%	6·276 −6·7%	6·478 −3·6%	σ_θ inner
3	1·9097	1·167 −38·9%	1·5406 −19·2%	1·735 −9·1%	1·826 −4·4%	σ_r inner
4	12·05	8·731 −27·5%	10·294 −14·6%	11·16 −7·3%	11·62 −3·5%	σ_r inner
5	0·924	0·8000 −13·4%	0·8586 −7%	0·8912 −3·4%	0·9088 −1·6%	σ_r inner
6	7·027	4·459 −36·5%	5·642 −19·7%	6·315 −10%	6·677 −5%	σ_r inner
6	3·4628	2·893 −16·5%	3·202 −7·5%	3·338 −3·7%	3·215 −7·3%	σ_r outer

Table 9.2. Circular flat plate: Maximum stresses and percentage errors

ciently fine mesh is employed. It is also clear that by the judicious use of mesh grading, improved accuracy can be obtained without increasing the total number of elements employed.

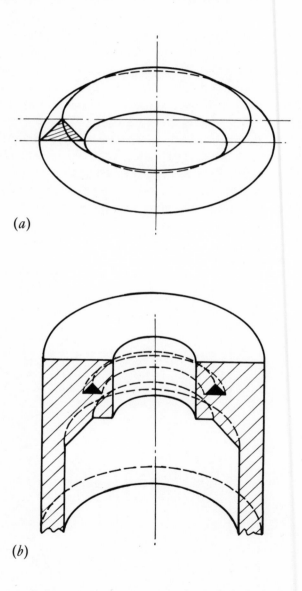

(a)

(b)

Fig. 9.10. *a* Typical axisymmetric element. *b* Axisymmetric solid

Solids of revolution

Axisymmetric solids and thick-walled bodies of revolution can be idealised by finite elements of revolution. Each element consists of a solid 'ring', the cross-section of which is the shape of the particular element chosen. Rectangularly-shaped, triangularly-shaped and quadrilaterally-shaped elements have been adopted. However, owing to its simplicity and versatility attention here is confined to the triangle of revolution. This element was initially reported by Clough and Rashid[5] and applied by Wilson[6] to the solution of a rocket nozzle subjected to thermal and pressure loading.

Figure 9.10 shows a section through an axially symmetric solid with a typical finite element consisting of three ring nodes. The derivation of the stiffness and stress matrices follows and it is seen that it is similar to that for the plane elasticity triangle developed in Chapter 5. The main difference lies in the number of stress com-components: an additional component, the hoop or tangential stress, has now to be included.

Step I: Choose suitable co-ordinate system and number nodes

It is convenient to use cylindrical polar co-ordinates (r, z) and the nodes are numbered 1, 2, 3 anti-clockwise (Figure 9.11). The nodal displacement and force vectors are written as

$$\{\delta_1\} = \begin{Bmatrix} u_1 \\ v_1 \end{Bmatrix} \qquad \{F_1\} = \begin{Bmatrix} F_{r1} \\ F_{z1} \end{Bmatrix}$$

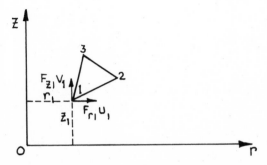

Fig. 9.11. Co-ordinates and node numbering for axisymmetric solid element

where F_r and u are in the radial direction, and F_z, v are in the axial direction.

The complete displacement and force vectors for the element may then be written as

$$\{\delta^e\} = \begin{Bmatrix} \{\delta_1\} \\ \{\delta_2\} \\ \{\delta_3\} \end{Bmatrix} = \begin{Bmatrix} u_1 \\ v_1 \\ u_2 \\ v_2 \\ u_3 \\ v_3 \end{Bmatrix} \tag{9.28}$$

$$\{F^e\} = \begin{Bmatrix} \{F_1\} \\ \{F_2\} \\ \{F_3\} \end{Bmatrix} = \begin{Bmatrix} F_{r1} \\ F_{z1} \\ F_{r2} \\ F_{z2} \\ F_{r3} \\ F_{z3} \end{Bmatrix} \tag{9.29}$$

Each of these vectors contains six terms and hence the element stiffness matrix $[K^e]$ is a 6×6 matrix.

$$\{F^e\} = [K^e]\{\delta^e\} \tag{I}$$

Step II: Choose displacement function $[f(r, z)]$ that defines displacement $\{\delta(r, z)\}$ at any point in element

For an axisymmetric solid the displacement of any point can be obtained by considering a radial and an axial displacement (u, v), there being no displacement in the hoop or tangential direction. This element, like the plane elasticity triangle in Chapter 5, has six degrees of freedom and therefore six unknown coefficients $(\alpha_1, \alpha_2, ..., \alpha_6)$ can be taken in the polynomial representing the permitted displacement pattern. The simplest representation is given by the two linear polynomials

$$\left.\begin{array}{l} u = \alpha_1 + \alpha_2 r + \alpha_3 z \\ v = \alpha_4 + \alpha_5 r + \alpha_6 z \end{array}\right\} \tag{9.30}$$

Since these are both linear in r and z, displacement continuity is ensured along the interface between adjoining elements.

Re-writing equation 9.30 in matrix form,

$$\{\delta(r, z)\} = \begin{Bmatrix} u \\ v \end{Bmatrix} = \begin{bmatrix} 1 & r & z & 0 & 0 & 0 \\ 0 & 0 & 0 & 1 & r & z \end{bmatrix} \begin{Bmatrix} \alpha_1 \\ \alpha_2 \\ \alpha_3 \\ \alpha_4 \\ \alpha_5 \\ \alpha_6 \end{Bmatrix} \tag{9.31}$$

or briefly

$$\{\delta(r, z)\} = [f(r, z)]\{\alpha\} \tag{II}$$

Step III: Express state of displacement within element $\{\delta(r, z)\}$ in terms of nodal displacements $\{\delta^e\}$

This step is achieved by substituting the values of the nodal co-ordinates into equation II and solving for $\{\alpha\}$. At node 1,

$$\{\delta_1\} = \{\delta(r_1, z_1)\} = [f(r_1, z_1)]\{\alpha\}$$

Substituting for $[f(r_1, z_1)]$ from equation 9.31,

$$\{\delta_1\} = \begin{bmatrix} 1 & r_1 & z_1 & 0 & 0 & 0 \\ 0 & 0 & 0 & 1 & r_1 & z_1 \end{bmatrix} \{\alpha\} \tag{9.32a}$$

Similarly for nodes 2 and 3,

$$\{\delta_2\} = \begin{bmatrix} 1 & r_2 & z_2 & 0 & 0 & 0 \\ 0 & 0 & 0 & 1 & r_2 & z_2 \end{bmatrix} \{\alpha\} \tag{9.32b}$$

$$\{\delta_3\} = \begin{bmatrix} 1 & r_3 & z_3 & 0 & 0 & 0 \\ 0 & 0 & 0 & 1 & r_3 & z_3 \end{bmatrix} \{\alpha\} \qquad (9.32c)$$

For the element as a whole these matrices can be combined to give

$$\{\delta^e\} = \begin{Bmatrix} \{\delta_1\} \\ \{\delta_2\} \\ \{\delta_3\} \end{Bmatrix} = \begin{bmatrix} 1 & r_1 & z_1 & 0 & 0 & 0 \\ 0 & 0 & 0 & 1 & r_1 & z_1 \\ 1 & r_2 & z_2 & 0 & 0 & 0 \\ 0 & 0 & 0 & 1 & r_2 & z_2 \\ 1 & r_3 & z_3 & 0 & 0 & 0 \\ 0 & 0 & 0 & 1 & r_3 & z_3 \end{bmatrix} \begin{Bmatrix} \alpha_1 \\ \alpha_2 \\ \alpha_3 \\ \alpha_4 \\ \alpha_5 \\ \alpha_6 \end{Bmatrix} \qquad (9.33)$$

which may be written briefly as

$$\{\delta^e\} = [A]\{\alpha\} \qquad (9.34)$$

The unknown coefficients $(\alpha_1, ..., \alpha_6)$ are now determined from equation 9.33 by inverting matrix $[A]$, to yield

$$\{\alpha\} = [A]^{-1}\{\delta^e\} \qquad (9.35)$$

The inverse of matrix $[A]$ is presented in equation 9.36, having been obtained in the same manner as the corresponding matrix in step III of Chapter 5. Thus a relationship between the unknown coefficients $\{\alpha\}$ and the nodal displacements $\{\delta^e\}$ has been obtained.

$$[A]^{-1} = \frac{1}{2\Delta} \times$$

$$\begin{bmatrix} r_2 z_3 - r_3 z_2 & 0 & -r_1 z_3 + r_3 z_1 & 0 & r_1 z_2 - r_2 z_1 & 0 \\ z_2 - z_3 & 0 & z_3 - z_1 & 0 & z_1 - z_2 & 0 \\ r_3 - r_2 & 0 & r_1 - r_3 & 0 & r_2 - r_1 & 0 \\ 0 & r_2 z_3 - r_3 z_2 & 0 & -r_1 z_3 + r_3 z_1 & 0 & r_1 z_2 - r_2 z_1 \\ 0 & z_2 - z_3 & 0 & z_3 - z_1 & 0 & z_1 - z_2 \\ 0 & r_3 - r_2 & 0 & r_1 - r_3 & 0 & r_2 - r_1 \end{bmatrix}$$

$$(9.36)$$

where

$$2\Delta = \det \begin{vmatrix} 1 & r_1 & z_1 \\ 1 & r_2 & z_2 \\ 1 & r_3 & z_3 \end{vmatrix} = 2 \times \text{area of triangle}$$

Using equation II the displacements $\{\delta(r, z)\}$ at any point in the element can now be determined in terms of the nodal displacements $\{\delta^e\}$.

$$\{\delta(r, z)\} = [f(r, z)]\{\alpha\}$$
$$\{\delta(r, z)\} = [f(r, z)][A]^{-1}\{\delta^e\} \qquad \text{(III)}$$

Step IV: Relate strains $\{\varepsilon(r, z)\}$ at any point in element to displacements $\{\delta(r, z)\}$ and hence to nodal displacements $\{\delta^e\}$

Using the theory of elasticity,[7] the four strain components for an axisymmetric solid are

$$\{\varepsilon(r, z)\} = \begin{Bmatrix} \varepsilon_r \\ \varepsilon_z \\ \varepsilon_\theta \\ \gamma_{rz} \end{Bmatrix} \qquad (9.37)$$

where ε_r (radial), ε_z (axial) and ε_θ (hoop) are the direct strains, and γ_{rz} is the shearing strain. Reference 7 gives the following relationships between strains ε and displacements u, v:

$$\left.\begin{aligned} \varepsilon_r &= \frac{\partial u}{\partial r}, \qquad \varepsilon_z = \frac{\partial v}{\partial z}, \\[2mm] \varepsilon_\theta &= \frac{u}{r}, \quad \text{and} \quad \gamma_{rz} = \frac{\partial u}{\partial z} + \frac{\partial v}{\partial r} \end{aligned}\right\} \qquad (9.38)$$

Substituting for u and v from equation 9.30,

$$\varepsilon_r = \alpha_2 \qquad \varepsilon_z = \alpha_6$$

$$\varepsilon_\theta = \frac{\alpha_1}{r} + \alpha_2 + \alpha_3 \frac{z}{r} \quad \text{and} \quad \gamma_{rz} = \alpha_3 + \alpha_5$$

Thus in matrix form,

$$\{\varepsilon(r, z)\} = \begin{Bmatrix} \varepsilon_r \\ \varepsilon_z \\ \varepsilon_\theta \\ \gamma_{rz} \end{Bmatrix} = \begin{bmatrix} 0 & 1 & 0 & 0 & 0 & 0 \\ 0 & 0 & 0 & 0 & 0 & 1 \\ \dfrac{1}{r} & 1 & \dfrac{z}{r} & 0 & 0 & 0 \\ 0 & 0 & 1 & 0 & 1 & 0 \end{bmatrix} \begin{Bmatrix} \alpha_1 \\ \alpha_2 \\ \alpha_3 \\ \alpha_4 \\ \alpha_5 \\ \alpha_6 \end{Bmatrix} \tag{9.39}$$

or simply

$$\{\varepsilon(r, z)\} = [C]\{\alpha\} \tag{9.40}$$

Substituting for $\{\alpha\}$ from equation 9.35,

$$\{\varepsilon(r, z)\} = [C][A]^{-1}\{\delta^e\} \tag{9.41}$$

or

$$\{\varepsilon(r, z)\} = [B]\{\delta^e\} \tag{IV}$$

where $[B] = [C][A]^{-1}$. Matrix $[B]$ is a 4×6 matrix and is presented in full in equation 9.42.

$$[B] = \frac{1}{2\Delta} \times$$

$$\begin{bmatrix} z_2 - z_3 & 0 & z_3 - z_1 & 0 & z_1 - z_2 & 0 \\ 0 & r_3 - r_2 & 0 & r_1 - r_3 & 0 & r_2 - r_1 \\ \dfrac{r_2 z_3 - r_3 z_2}{r} & \dfrac{-r_1 z_3 + r_3 z_1}{r} & \dfrac{r_1 z_2 - r_2 z_1}{r} & & & \\ +(z_2 - z_3) & 0 & +(z_3 - z_1) & 0 & +(z_1 - z_2) & 0 \\ +\dfrac{z}{r}(r_3 - r_2) & & +\dfrac{z}{r}(r_1 - r_3) & & +\dfrac{z}{r}(r_2 - r_1) & \\ r_3 - r_2 & z_2 - z_3 & r_1 - r_3 & z_3 - z_1 & r_2 - r_1 & z_1 - z_2 \end{bmatrix}$$

$$\tag{9.42}$$

where

$$2\Delta = \det \begin{vmatrix} 1 & r_1 & z_1 \\ 1 & r_2 & z_2 \\ 1 & r_3 & z_3 \end{vmatrix}$$

Step V: Relate internal stresses $\{\sigma(r, z)\}$ to strains $\{\varepsilon(r, z)\}$ and to nodal displacements

The stress components for axisymmetric solids are

$$\{\sigma(r, z)\} = \begin{Bmatrix} \sigma_r \\ \sigma_z \\ \sigma_\theta \\ \tau_{rz} \end{Bmatrix} \tag{9.43}$$

where σ_r (radial), σ_z (axial) and σ_θ (hoop) are the direct stresses and τ_{rz} is the shear stress.

Elastic theory[7] indicates that the strains are related to the stresses as follows.

$$\left. \begin{aligned} \varepsilon_r &= \frac{\sigma_r}{E} - v\frac{\sigma_z}{E} - v\frac{\sigma_\theta}{E} \\[1mm] \varepsilon_z &= -v\frac{\sigma_r}{E} + \frac{\sigma_z}{E} - v\frac{\sigma_\theta}{E} \\[1mm] \varepsilon_\theta &= -v\frac{\sigma_r}{E} - v\frac{\sigma_z}{E} + \frac{\sigma_\theta}{E} \\[1mm] \gamma_{rz} &= \frac{\tau_{rz}}{G} \end{aligned} \right\} \tag{9.44}$$

where E is the elastic modulus, G is the modulus of rigidity and v is Poisson's ratio. Noting that $E = 2G(1+v)$, equation 9.44 can be solved to give the stresses in terms of the strains in equation 9.45.

$$\begin{Bmatrix} \sigma_r \\ \sigma_z \\ \sigma_\theta \\ \tau_{rz} \end{Bmatrix} = \frac{E(1-v)}{(1+v)(1-2v)} \begin{bmatrix} 1 & \dfrac{v}{1-v} & \dfrac{v}{1-v} & 0 \\[2mm] \dfrac{v}{1-v} & 1 & \dfrac{v}{1-v} & 0 \\[2mm] \dfrac{v}{1-v} & \dfrac{v}{1-v} & 1 & 0 \\[2mm] 0 & 0 & 0 & \dfrac{1-2v}{2(1-v)} \end{bmatrix} \begin{Bmatrix} \varepsilon_r \\ \varepsilon_z \\ \varepsilon_\theta \\ \gamma_{rz} \end{Bmatrix} \tag{9.45}$$

or briefly as

$$\{\sigma(r, z)\} = [D]\{\varepsilon(r, z)\} \tag{9.46}$$

where $[D]$ is the elasticity matrix for this particular case.

Substituting in equation 9.46 the expression for $\{\varepsilon(r, z)\}$ from equation IV, the stress–nodal displacement relationship is

$$\{\sigma(r, z)\} = [D][B]\{\delta^e\} \tag{V}$$

Step VI: Replace internal stresses $\{\sigma(r, z)\}$ with statically equivalent nodal forces $\{F^e\}$, relate nodal forces to nodal displacements $\{\delta^e\}$ and hence obtain element stiffness matrix

$$\{F^e\} = [\textstyle\int [B]^T[D][B] \, d(\text{vol})]\{\delta^e\} \tag{VI}$$

where

$$[\textstyle\int [B]^T[D][B] \, d(\text{vol})]$$

is the element stiffness matrix $[K^e]$. The integral is taken over the volume of the element, and for a body of revolution,

$$d(\text{vol}) = 2\pi r \, dr \, dz$$

Hence

$$[K^e] = \textstyle\iint [B]^T[D][B] 2\pi r \, dr \, dz \tag{9.47}$$

The $[B]$ and $[D]$ matrices for this case have already been obtained. Unlike the case of plane elasticity, matrix $[B]$ (equation 9.42) now depends upon the co-ordinates r and z, and hence the matrix operations indicated in equation 9.47 must be performed and each term integrated with respect to both r and z, in order to obtain $[K^e]$. To avoid this lengthy process, a simple approximation has been used that is found to give good results. This involves evaluating the $[B]$ matrix for a centroidal point within the element defined by the co-ordinates $(\bar{r}, \bar{z})$, where $\bar{r} = \frac{1}{3}(r_1 + r_2 + r_3)$ and $\bar{z} = \frac{1}{3}(z_1 + z_2 + z_3)$. Then the stiffness matrix becomes simply

$$[K^e] = 2\pi [B]^T[D][B]\bar{r}\Delta \tag{9.48}$$

where Δ is the area of the triangle. In matrix $[B]$, the values $\bar{r}$ and $\bar{z}$ are substituted for r and z, which then renders the matrix constant for any particular element.

The triple product $[B]^T[D][B]$ in equation 9.48 has not been carried out algebraically as was done in the case of the plane elasticity problems in Chapter 5.

Step VII: Establish stress–displacement matrix $[H]$

The final step is the determination of the stresses at any point in the element in terms of the nodal displacements of the element. The relationship given in equation V enables this to be done.

$$\{\sigma(r, z)\} = [D][B]\{\delta^e\} \tag{V}$$

or

$$\{\sigma(r, z)\} = [H]\{\delta^e\} \tag{VII}$$

As in step VI, if the centroidal values $(\bar{r}, \bar{z})$ are substituted in matrix $[B]$ the four stress components at the centroid of the element are obtained.

Application of axisymmetric solid triangular element

Example: Thick cylinder under internal pressure

An example of the use of this element can be illustrated by considering a thick cylinder under internal pressure, as shown in Figure 9.12. The three finite element idealisations which have been employed are shown in Figure 9.13. Figures 9.14 and 9.15 present the results for displacements and stresses in the cylinder. The full lines show the theoretical values[7] and the points represent the finite element values for the various idealisations. The graphs show that even for the coarsest mesh, the maximum radial deflection is within 0·05% of the theoretical value, while the radial and hoop stresses are within 4%. For the finest mesh employed the stresses are now seen to be within 0·5% of the exact values.

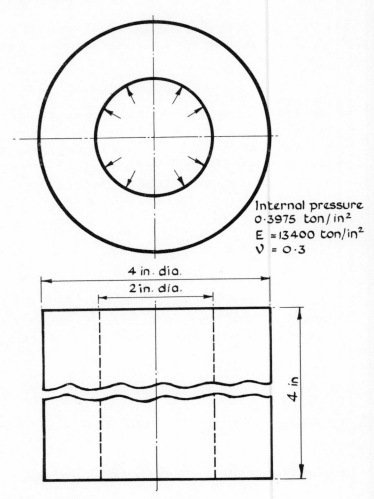

Internal pressure
0.3975 ton/in²
E = 13400 ton/in²
ν = 0.3

Fig. 9.12. Thick cylinder under internal pressure

Conclusions

For both the axisymmetric shell and solid elements the simplest displacement functions have been employed, and it is seen that accurate answers are obtained provided suitable idealisations of the actual structure are used. For a detailed discussion of the selection of displacement functions for these elements and the effects upon accuracy and simplicity in use, the reader is referred to references 2 and 8.

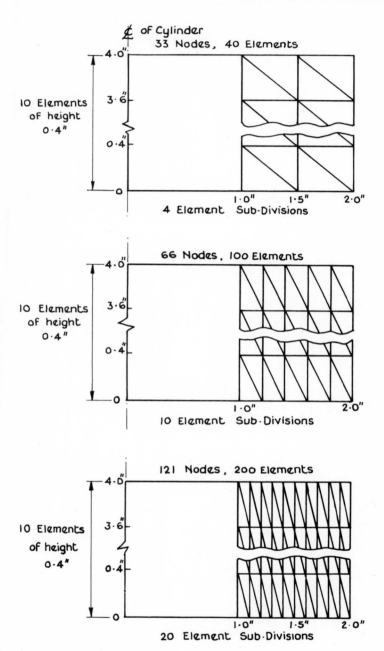

Fig. 9.13. Idealisations of thick cylinder

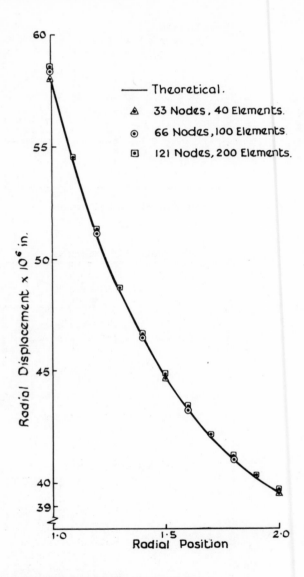

Fig. 9.14. Radial displacement in thick cylinder

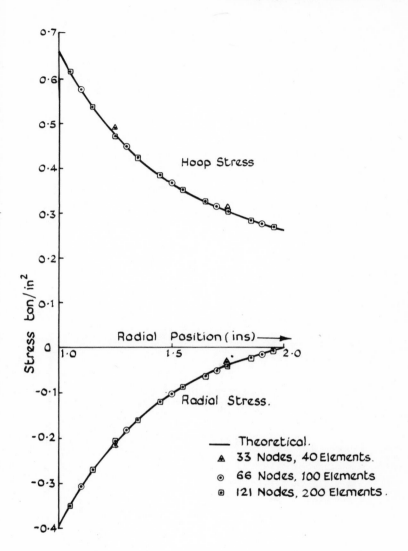

Fig. 9.15. Radial and hoop stresses in thick cylinder

References

1. GRAFTON, P. E. and STROME, D. R. Analysis of axisymmetric shells by the direct stiffness method. *AIAA Journal*, Vol. 1, No. 10. 1963, pp. 2342–2347.
2. PERCY, J. H., PIAN, T. H. H., KLEIN, S. and NAVARATNA, D. R. Application of matrix displacement method to linear elastic analysis of shells of revolution. *AIAA Journal*, Vol. 3, No. 11, 1965, pp. 2138–2145.
3. TIMOSHENKO, S. P. and WOINOWSKY-KRIEGER, S. *Theory of plates and shells*. New York, McGraw-Hill Book Co., 1959.
4. ROARK, R. J. *Formulas for stress and strain*. New York, McGraw-Hill Publishing Co., 1965.
5. CLOUGH, R. W. and RASHID, Y. Finite element analysis of axisymmetric solids. *Proceedings of the American Society of Civil Engineers/EM Division*, Vol. 91, 1965, pp. 71–85.
6. WILSON, E. L. Structural analysis of axisymmetric solids. *AIAA Journal*, Vol. 3, No. 12, 1965, pp. 2269–2274.
7. TIMOSHENKO, S. P. and GOODIER, J. N. *Theory of elasticity*. New York, McGraw-Hill Book Co., 1951.
8. CLOUGH, R. W. The finite element method in structural mechanics. Chapter 7 of *Stress analysis* edited by O. C. Zieneiwicz and G. Holister. London, John Wiley & Sons Ltd, 1965.

CHAPTER TEN

Programming

Introduction

In Chapters 3 to 9 the steps necessary to obtain the element stiffness matrices $[K^e]$ and stress matrices $[H]$ have been developed. A finite element solution involves calculating the stiffness matrices for every element in the idealised structure and then assembling the overall structural stiffness matrix $[K]$ for the complete structure.

In Chapter 2 it has been shown that the loading as expressed by the vector $\{F\}$ is related to the displacement vector $\{\delta\}$ by equation 10.1 where $[K]$ is the overall stiffness matrix for the complete structure.

$$\{F\} = [K]\{\delta\} \qquad (10.1)$$

In the stiffness method of analysis the unknown quantities are the displacements and so the object of the solution procedure is to determine the nodal displacements $\{\delta\}$ in equation 10.1. Once these are known, the element stresses can be calculated using the element stress matrix $[H]$ (step VII).

The numerical work involved in performing this analysis for any real structure makes an electronic computer an essential adjunct to the application of the finite element method. In this chapter a number of the techniques which are adopted when developing a suitable computer program are discussed. Computer programs can be built up in a number of parts called subroutines, each of which carries out a particular function in the overall analysis. For example, the evaluation of the element stiffness matrix $[K^e]$ for a particular shape of element and type of problem is normally done in a subroutine written specifically for that shape of element and type of problem, whether it is one of plane stress, plane strain or plate flexure.

The structural idealisation (or subdivision into finite elements) has been discussed for both linear structures and continua. It is seen that the basic steps in determining the element stiffness and stress matrices are the same for these various elements. Similarly, the solution of the equations for nodal displacements and element stresses follows a pattern that is independent of the type of element used. This enables a finite element computer system to be developed in which some parts (subroutines) of the program are completely general. An obvious example is the subroutine that has to be written for solving the simultaneous equations represented in equation 10.1. Clearly it is also necessary to have a main routine that controls the order and use of the various subroutines.

The development of a typical computer program is now discussed in some detail.

Computer analysis

Table 10.1 summarises the basic requirements of the computer program necessary for the complete solution of a problem by the finite element method. The analysis is carried out using input data which describe fully the idealised structure and its loading, and produces output consisting of tabulated nodal displacements and element stresses. Each of the steps in Table 10.1 is now described separately.

Input data

To specify the problem it is necessary to provide the computer with input data. This material consists of data specifying the geometry of the idealised structure, its material properties and the way it is

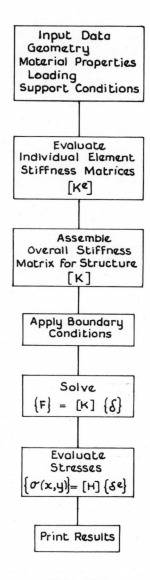

Table 10.1

loaded and supported in space. The data also include certain control numbers that may help the generality and efficiency of the program and should be supplied early in the input data, such as the total number of nodes and elements or a control number which indicates whether the problem is plane stress, plane strain or plate flexure. This enables the main routine to ascertain how much storage is required and which subroutines are needed in the analysis.

The following details are needed specifically.

 i The number and the co-ordinates of the nodal points; it is convenient to number the nodal points $1...i...n$, so that for each node the node number i and its co-ordinates (x_i, y_i) are supplied. The importance of adopting the most effective method of numbering the nodes has already been demonstrated.
 ii The finite elements may also be numbered $1...N$, so that for each element its nodal numbers (connectivity) and the corresponding nodal co-ordinates of the element can be supplied, thus enabling the dimensions of the element and its position in the overall structure to be determined. It is also necessary to specify the material properties of the element, i.e. Young's Modulus E and Poisson's ratio v.
 iii The applied loads.
 iv The 'fixed nodes'—corresponding to the various types of supports.

This no doubt appears a complicated procedure. However, it will soon be clear that it can easily be accomplished.

Consider a simple example, a deep cantilever carrying a load of 10 kN as shown in Figure 10.1. The cantilever is assumed fully fixed at the left-hand edge. For the purpose of this illustration the

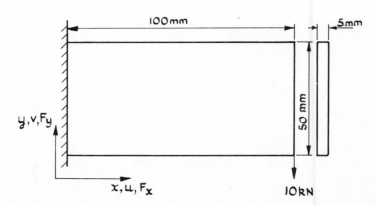

Fig. 10.1. Deep cantilever

structure is idealised as shown in Figure 10.2, although such a coarse idealisation would yield an inaccurate result. The idealisation gives the following information (taking the origin of the co-ordinate system at node 1).

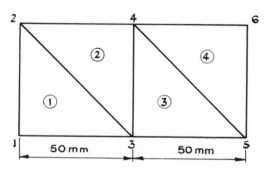

Fig. 10.2. Coarse finite element idealisation

   i Number of nodes: 6.
  ii Number of elements: 4.
 iii Type of element: triangular plane stress, uniform thickness.
  iv All elements have the same material properties throughout.
   v At each of the six nodes there are two degrees of freedom. For example, at node 3 there is a horizontal deflection component u_3 and a vertical deflection component v_3. The boundary conditions may be specified as follows:

$$u_1 = v_1 = u_2 = v_2 = 0$$

 vi The nodal co-ordinates and applied loading may be written as

Node	x-co-ordinate	y-co-ordinate	F_x	F_y
1	0	0	—	—
2	0	50	—	—
3	50	0	0	0
4	50	50	0	0
5	100	0	0	−10
6	100	50	0	0

Remember that if the loading at a joint is known, then the displacements of that joint are not known and vice versa. The forces F_x and F_y refer to externally applied loads.

vii Numbering the element nodes in an anti-clockwise manner as in Chapter 5, the element connectivity and material properties may be written as

Element	Connectivity			E	ν	t
1	1	3	2	200	0·3	5
2	2	3	4	200	0·3	5
3	3	5	4	200	0·3	5
4	4	5	6	200	0·3	5

Data cards may also be included to give the type of element being used in this case and the number of nodes and degrees of freedom so that the storage may be calculated.

The data can be input by means of punched cards or paper tape. Some of the input data, e.g. element nodal connectivity, may even be generated within the computer, provided suitable program instructions have been included.

Element stiffness matrix $[K^e]$

The computer program may contain a subroutine for each of the many types of finite elements available or it may only contain those needed for the problem at hand. Thus the main routine may or may not require a control number to indicate the type of element stiffness matrix $[K^e]$ called for in a particular problem. In many problems one type of element is used throughout and so it is unnecessary to retain subroutines for other elements which are not required in the analysis. However, to retain greater flexibility at the expense of greater complexity, it may be desirable to retain in the system all the various subroutines available.

The subroutine in which the element stiffness matrix $[K^e]$ is calculated requires certain information which is included in the input data—the nodal co-ordinates and element properties—and as soon as this has been read in by the main routine, $[K^e]$ can be calculated for each element in turn. The subroutine may contain the algebraic expression for each term k_{ij} of $[K^e]$ or it may start with the expression for $[B]$ and $[D]$ and evaluate these and then perform the matrix multiplication necessary to obtain the terms of $[K^e]$ as indicated in equation VI. For more complex elements it is often necessary to use numerical integration in order to obtain $[K^e]$. The details of this subroutine depend upon the type of element involved.

Assembly of structural stiffness $[K]$

As soon as each element stiffness matrix $[K^e]$ has been established, it is necessary to assemble this into the overall structural stiffness matrix $[K]$. The manner in which this is done can be illustrated by a simple example. Later, means of achieving more efficient use of the computer storage are shown.

Consider again the deep cantilever shown in Figure 10.1. The idealisation shown in Figure 10.2, with six nodes and four triangular elements, is again adopted. Since there are two degrees of freedom per node (u and v) this idealised structure has a total of twelve degrees of freedom, so that the overall structural stiffness matrix $[K]$ is a 12×12 matrix. For each triangular element $[K^e]$ is a 6×6 matrix (see Chapter 5).

The overall structural stiffness matrix $[K]$ is now formed. In equation 10.1,

$$\{F\} = [K]\{\delta\} \tag{10.1}$$

The nodal displacement vector and the nodal load vector consist of twelve terms which are now written in full in equation 10.2.

$$
\begin{Bmatrix}
F_{x1} \\
F_{y1} \\
F_{x2} \\
F_{y2} \\
F_{x3} \\
F_{y3} \\
F_{x4} \\
F_{y4} \\
F_{x5} \\
F_{y5} \\
F_{x6} \\
F_{y6}
\end{Bmatrix}
=
\begin{bmatrix}
& & \\
& \text{Overall } 12 \times 12 & \\
& \text{stiffness} & \\
& \text{matrix } [K] & \\
& & \\
& & \\
& & \\
\end{bmatrix}
\begin{Bmatrix}
u_1 \\
v_1 \\
u_2 \\
v_2 \\
u_3 \\
v_3 \\
u_4 \\
v_4 \\
u_5 \\
v_5 \\
u_6 \\
v_6
\end{Bmatrix}
\tag{10.2}
$$

This overall structural stiffness matrix $[K]$ is of size 12×12 and consists of the sum of the contributions of the individual element stiffness matrices $[K^e]$.

Consider the first element (Figure 10.3), where the node numbers correspond to those for the overall structure. In Chapter 5, step I, when developing the element stiffness matrix $[K^e]$ the convention of numbering the nodes of each element in an anti-clockwise manner was adopted. Thus for element 1 in Figure 10.3, to allow for the difference between the 'element' and 'structure' node numbering systems, the nodal displacement vector $\{\delta^i\}$ and force vector $\{F^i\}$ must be written in the order shown in equation 10.3. The terms in the stiffness matrix correspond to the terms in equation 5.30 in Chapter 5.

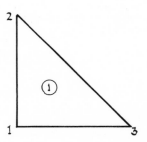

Fig. 10.3. Element 1 with overall structure node numbering

$$\{F^i\} = \begin{Bmatrix} F_{x1} \\ F_{y1} \\ F_{x3} \\ F_{y3} \\ F_{x2} \\ F_{y2} \end{Bmatrix} = \begin{bmatrix} k^i_{11} & k^i_{12} & k^i_{13} & k^i_{14} & k^i_{15} & k^i_{16} \\ k^i_{21} & k^i_{22} & k^i_{23} & k^i_{24} & k^i_{25} & k^i_{26} \\ k^i_{31} & k^i_{32} & k^i_{33} & k^i_{34} & k^i_{35} & k^i_{36} \\ k^i_{41} & k^i_{42} & k^i_{43} & k^i_{44} & k^i_{45} & k^i_{46} \\ k^i_{51} & k^i_{52} & k^i_{53} & k^i_{54} & k^i_{55} & k^i_{56} \\ k^i_{61} & k^i_{62} & k^i_{63} & k^i_{64} & k^i_{65} & k^i_{66} \end{bmatrix} \begin{Bmatrix} u_1 \\ v_1 \\ u_3 \\ v_3 \\ u_2 \\ v_2 \end{Bmatrix} \qquad (10.3)$$

Where the element stiffness matrix for element 1 is designated $[K^i]$, its terms are designated k^i_{ij}, and the nodal force vector by $\{F^i\}$ and the displacement vector by $\{\delta^i\}$. When the values of k_{ij} have been calculated (from equation 5.30) they must be assembled into the $[K]$ matrix of equation 10.2 and this assembly has been done and is shown in equation 10.4. It is very important to note the difference in order of the terms between equations 10.3 and 10.4.

Special attention should be given at all times to ensure that the terms as calculated in the individual element procedures are inserted into the correct location in the overall matrix.

$$
\begin{Bmatrix}
F_{x1} \\ F_{y1} \\ F_{x2} \\ F_{y2} \\ F_{x3} \\ F_{y3} \\ F_{x4} \\ F_{y4} \\ F_{x5} \\ F_{y5} \\ F_{x6} \\ F_{y6}
\end{Bmatrix}
=
\begin{bmatrix}
k^i_{11} & k^i_{12} & k^i_{15} & k^i_{16} & k^i_{13} & k^i_{14} & & & & & & \\
k^i_{21} & k^i_{22} & k^i_{25} & k^i_{26} & k^i_{23} & k^i_{24} & & & & & & \\
k^i_{51} & k^i_{52} & k^i_{55} & k^i_{56} & k^i_{53} & k^i_{54} & & & & & & \\
k^i_{61} & k^i_{62} & k^i_{65} & k^i_{66} & k^i_{63} & k^i_{64} & & & & & & \\
k^i_{31} & k^i_{32} & k^i_{35} & k^i_{36} & k^i_{33} & k^i_{34} & & & & & & \\
k^i_{41} & k^i_{42} & k^i_{45} & k^i_{46} & k^i_{43} & k^i_{44} & & & & & & \\
 & & & & & & & & & & & \\
 & & & & & & & & & & & \\
 & & & & & & & & & & & \\
 & & & & & & & & & & & \\
 & & & & & & & & & & & \\
 & & & & & & & & & & & \\
\end{bmatrix}
\begin{Bmatrix}
u_1 \\ v_1 \\ u_2 \\ v_2 \\ u_3 \\ v_3 \\ u_4 \\ v_4 \\ u_5 \\ v_5 \\ u_6 \\ v_6
\end{Bmatrix}
$$

$$(10.4)$$

For element 2 the node numbers are as shown in Figure 10.4, and the load and displacement vectors and the element stiffness matrix for element 2 are given in equation 10.5.

$$
\{F^{ii}\} =
\begin{Bmatrix}
F_{x2} \\ F_{y2} \\ F_{x3} \\ F_{y3} \\ F_{x4} \\ F_{y4}
\end{Bmatrix}
=
\begin{bmatrix}
k^{ii}_{11} & k^{ii}_{12} & k^{ii}_{13} & k^{ii}_{14} & k^{ii}_{15} & k^{ii}_{16} \\
k^{ii}_{21} & k^{ii}_{22} & k^{ii}_{23} & k^{ii}_{24} & k^{ii}_{25} & k^{ii}_{26} \\
k^{ii}_{31} & k^{ii}_{32} & k^{ii}_{33} & k^{ii}_{34} & k^{ii}_{35} & k^{ii}_{36} \\
k^{ii}_{41} & k^{ii}_{42} & k^{ii}_{43} & k^{ii}_{44} & k^{ii}_{45} & k^{ii}_{46} \\
k^{ii}_{51} & k^{ii}_{52} & k^{ii}_{53} & k^{ii}_{54} & k^{ii}_{55} & k^{ii}_{56} \\
k^{ii}_{61} & k^{ii}_{62} & k^{ii}_{63} & k^{ii}_{64} & k^{ii}_{65} & k^{ii}_{66}
\end{bmatrix}
\begin{Bmatrix}
u_2 \\ v_2 \\ u_3 \\ v_3 \\ u_4 \\ v_4
\end{Bmatrix}
\quad (10.5)
$$

Assembling these terms into equation 10.2, equation 10.6 is obtained.

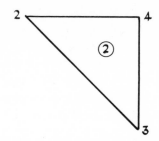

Fig. 10.4. Element 2 with overall structure node numbering

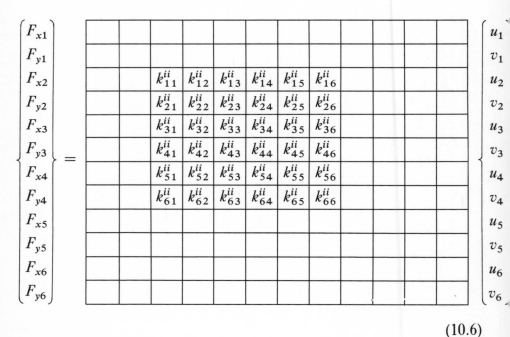

$$(10.6)$$

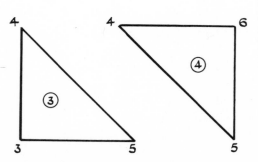

Fig. 10.5. Node numbering for elements 3 and 4

Repeating the procedure for elements 3 and 4 (Figure 10.5), the corresponding nodal load and nodal displacement equations are found to be as given in equations 10.7 and 10.8.

$$\{F^{iii}\} = \begin{Bmatrix} F_{x3} \\ F_{y3} \\ F_{x5} \\ F_{y5} \\ F_{x4} \\ F_{y4} \end{Bmatrix} = \begin{bmatrix} k_{11}^{iii} & k_{12}^{iii} & k_{13}^{iii} & - & - & k_{16}^{iii} \\ k_{21}^{iii} & k_{22}^{iii} & - & - & - & k_{26}^{iii} \\ k_{31}^{iii} & - & - & - & - & k_{36}^{iii} \\ k_{41}^{iii} & - & - & - & - & k_{46}^{iii} \\ k_{51}^{iii} & - & - & - & - & k_{56}^{iii} \\ k_{61}^{iii} & - & - & - & - & k_{66}^{iii} \end{bmatrix} \begin{Bmatrix} u_3 \\ v_3 \\ u_5 \\ v_5 \\ u_4 \\ v_4 \end{Bmatrix} \quad (10.7)$$

$$\{F^{iv}\} = \begin{Bmatrix} F_{x4} \\ F_{y4} \\ F_{x5} \\ F_{y5} \\ F_{x6} \\ F_{y6} \end{Bmatrix} = \begin{bmatrix} k_{11}^{iv} & k_{12}^{iv} & k_{13}^{iv} & - & - & k_{16}^{iv} \\ k_{21}^{iv} & k_{22}^{iv} & - & - & - & k_{26}^{iv} \\ k_{31}^{iv} & - & - & - & - & k_{36}^{iv} \\ k_{41}^{iv} & - & - & - & - & k_{46}^{iv} \\ k_{51}^{iv} & - & - & - & - & k_{56}^{iv} \\ k_{61}^{iv} & - & - & - & - & k_{66}^{iv} \end{bmatrix} \begin{Bmatrix} u_4 \\ v_4 \\ u_5 \\ v_5 \\ u_6 \\ v_6 \end{Bmatrix} \quad (10.8)$$

When the stiffness terms for all four elements are assembled into equation 10.2, equation 10.9 (p. 188) is obtained.

Equation 10.9 shows how the overall structural stiffness matrix has been assembled by inserting all the element stiffness matrices $[K^e]$ into the appropriate locations. It was noted in equation 5.30 that $[K^e]$ is symmetrical, i.e. terms such as $k_{12} = k_{21}$ etc. Similarly it is clear that $[K]$ is symmetrical in equation 10.9. In addition, it is also clear that terms in the bottom left-hand and top right-hand corners of $[K]$ are all zero, the non-zero terms all lying within a band on each side of the leading diagonal. In order to economise on the computer storage space, use is made both of the symmetry and of the banded nature of $[K]$. This point is discussed later.

$$
\begin{Bmatrix}
F_{x1} \\ F_{y1} \\ F_{x2} \\ F_{y2} \\ F_{x3} \\ F_{y3} \\ F_{x4} \\ F_{y4} \\ F_{x5} \\ F_{y5} \\ F_{x6} \\ F_{y6}
\end{Bmatrix}
=
\begin{bmatrix}
k^{i}_{11} & k^{i}_{12} & k^{i}_{15} & k^{i}_{16} & k^{i}_{13} & k^{i}_{14} & & & & & & \\[4pt]
k^{i}_{21} & k^{i}_{22} & k^{i}_{25} & k^{i}_{26} & k^{i}_{23} & k^{i}_{24} & & & & & & \\[4pt]
k^{i}_{51} & k^{i}_{52} & k^{i}_{55}+k^{ii}_{11} & k^{i}_{56}+k^{ii}_{12} & k^{i}_{53}+k^{ii}_{13} & k^{i}_{54}+k^{ii}_{14} & k^{ii}_{15} & k^{ii}_{16} & & & & \\[4pt]
k^{i}_{61} & k^{i}_{62} & k^{i}_{65}+k^{ii}_{21} & k^{i}_{66}+k^{ii}_{22} & k^{i}_{63}+k^{ii}_{23} & k^{i}_{64}+k^{ii}_{24} & k^{ii}_{25} & k^{ii}_{26} & & & & \\[4pt]
k^{i}_{31} & k^{i}_{32} & k^{i}_{35}+k^{ii}_{31} & k^{i}_{36}+k^{ii}_{32} & k^{i}_{33}+k^{ii}_{33}+k^{iii}_{11} & k^{i}_{34}+k^{ii}_{34}+k^{iii}_{12} & k^{ii}_{35}+k^{iii}_{15} & k^{ii}_{36}+k^{iii}_{16} & k^{iii}_{13} & k^{iii}_{14} & & \\[4pt]
k^{i}_{41} & k^{i}_{42} & k^{i}_{45}+k^{ii}_{41} & k^{i}_{46}+k^{ii}_{42} & k^{i}_{43}+k^{ii}_{43}+k^{iii}_{21} & k^{i}_{44}+k^{ii}_{44}+k^{iii}_{22} & k^{ii}_{45}+k^{iii}_{25} & k^{ii}_{46}+k^{iii}_{26} & k^{iii}_{23} & k^{iii}_{24} & & \\[4pt]
& & k^{ii}_{51} & k^{ii}_{52} & k^{ii}_{53}+k^{iii}_{51} & k^{ii}_{54}+k^{iii}_{52} & k^{ii}_{55}+k^{iii}_{55}+k^{iv}_{11} & k^{ii}_{56}+k^{iii}_{56}+k^{iv}_{12} & k^{iii}_{53}+k^{iv}_{13} & k^{iii}_{54}+k^{iv}_{14} & k^{iv}_{15} & k^{iv}_{16} \\[4pt]
& & k^{ii}_{61} & k^{ii}_{62} & k^{ii}_{63}+k^{iii}_{61} & k^{ii}_{64}+k^{iii}_{62} & k^{ii}_{65}+k^{iii}_{65}+k^{iv}_{21} & k^{ii}_{66}+k^{iii}_{66}+k^{iv}_{22} & k^{iii}_{63}+k^{iv}_{23} & k^{iii}_{64}+k^{iv}_{24} & k^{iv}_{25} & k^{iv}_{26} \\[4pt]
& & & & k^{iii}_{31} & k^{iii}_{32} & k^{iii}_{35}+k^{iv}_{31} & k^{iii}_{36}+k^{iv}_{32} & k^{iii}_{33}+k^{iv}_{33} & k^{iii}_{34}+k^{iv}_{34} & k^{iv}_{35} & k^{iv}_{36} \\[4pt]
& & & & k^{iii}_{41} & k^{iii}_{42} & k^{iii}_{45}+k^{iv}_{41} & k^{iii}_{46}+k^{iv}_{42} & k^{iii}_{43}+k^{iv}_{43} & k^{iii}_{44}+k^{iv}_{44} & k^{iv}_{45} & k^{iv}_{46} \\[4pt]
& & & & & & k^{iv}_{51} & k^{iv}_{52} & k^{iv}_{53} & k^{iv}_{54} & k^{iv}_{55} & k^{iv}_{56} \\[4pt]
& & & & & & k^{iv}_{61} & k^{iv}_{62} & k^{iv}_{63} & k^{iv}_{64} & k^{iv}_{65} & k^{iv}_{66}
\end{bmatrix}
\begin{Bmatrix}
u_1 \\ v_1 \\ u_2 \\ v_2 \\ u_3 \\ v_3 \\ u_4 \\ v_4 \\ u_5 \\ v_5 \\ u_6 \\ v_6
\end{Bmatrix}
\tag{10.9}
$$

Application of boundary conditions

In Chapter 2 it was shown how important it is to ensure that the structure is adequately supported before proceeding to the solution stage, failure to do so rendering the problem insoluble. In general, the structure is supported at a number of nodes such that various displacements are prevented. In such cases, as seen in Chapter 2, it is necessary to delete the appropriate row and column from the overall stiffness matrix $[K]$. It is, therefore, necessary that the program contains a routine in which the fixed displacements are specified and the matrix $[K]$ is modified accordingly. Several different methods of actually performing this operation exist. Some alternatives are as follows.

i Multiply the diagonal terms of $[K]$ corresponding to the fixed displacements, i.e. k_{ii}, by a large number such as 10^{20}. For example, assume that the x displacement at node i is to be zero, i.e. $u_i = 0$ is required. Then writing out equation 10.1 for row i,

$$F_{xi} = k_{i1}u_1 + k_{i2}v_1 + k_{i3}u_2 + \cdots + k_{ii}u_i + \cdots + k_{in}v_n$$

Then multiplying the diagonal term k_{ii} by 10^{20},

$$F_{xi} = k_{i1}u_1 + k_{i2}v_1 + \cdots + (10^{20})k_{ii}u_i + \cdots + k_{in}v_n,$$

and thus

$$u_i = \frac{F_{xi} - (k_{i1}u_1 + k_{i2}v_1 + \cdots + k_{in}v_n)}{(10^{20})k_{ii}}$$

Since the diagonal term is now so much greater than the off-diagonal terms, the value of the displacement u_i is very close to zero.

ii Replace the diagonal terms of $[K]$ corresponding to the fixed displacement by unity and replace the rest of the terms in the corresponding row and column by zeros.

iii Eliminate the row and column corresponding to the fixed displacement.

iv Allow for the fixed displacements in the assembly of $[K]$, so that the corresponding terms are not assembled.

The first two methods are essentially similar because they make use of the presence of a diagonal term that is large relative to the other terms in the same row and column. Similarly, the third and fourth methods give precisely the same result, but whereas the third requires a certain 'shuffling' of the matrix the fourth is more difficult actually to program. However, all four methods have the same effect, the influence of the prescribed zero displacement being eliminated.

The presence of elastic supports, e.g. non–rigid column supports in bridge problems, may be allowed for simply by adding the stiffness of the support to the diagonal term of the overall stiffness matrix corresponding to the appropriate displacement, provided that the corresponding term in the force vector is set to zero.

If certain displacements are prescribed as having known non–zero values then this effect too may be allowed for using methods similar to those described in sections (i) and (ii) above for treating zero displacements. For example, using method (i), assume that node i has a prescribed displacement d in the x direction. Then the diagonal term k_{ii} is multiplied by a large number e.g. (10^{20}). In addition, the corresponding term F_{xi} in the load vector is replaced by the diagonal term multiplied by the prescribed displacement. Thus writing out the equation for row i,

$$(10^{20})k_{ii}d = k_{i1}u_1 + k_{i2}v_1 + \cdots 10^{20}k_{ii}u_i + \cdots + k_{in}v_n$$

Consequently

$$u_i = \frac{(10^{20})k_{ii}d - \{k_{i1}u_1 + k_{i2}v_1 + \cdots + k_{in}v_n\}}{(10^{20})k_{ii}}$$

Since the diagonal term $(10^{20})k_{ii}$ is so much larger than the off-diagonal terms $k_{ii}u_1$, etc., then to a good approximation $u_i = d$.

Alternatively if method (ii) is used the term in the force vector is then simply replaced by the value of the prescribed displacement d. The case of a prescribed zero displacement is therefore a particular case of this more general process.

Solution for displacements

For all but the most trivial of problems, the final set of equations will be quite large. Several schemes for solving large systems of equations exist and among those that have successfully been used in conjunction with the finite element method are Gaussian elimination,[1] and the Gauss-Seidel[1] iterative and Cholesky square-root[1,2] methods.

It is this stage, together with the stage during which the overall stiffness matrix $[K]$ is assembled, which usually occupies most of the program time and therefore it is in these two stages that the ingenious programmer has most opportunity to exercise his talents. Consequently, later in this chapter each of these stages is discussed in more detail with particular reference to program efficiency. For the present, it is sufficient merely to think in terms of the program proceeding to a routine in which the final set of equations corresponding to equation 10.1 are solved for the displacements. Indeed, in many instances it may be possible for the engineer to obtain from a program library a standard subroutine to perform this task, thus relieving himself of the task of writing one.

Solution for stresses

Once the nodal displacements have been obtained the stresses within the element may be obtained quite simply. In the stress routine it is, first of all, necessary to extract the nodal displacements for the particular element in question from the complete set of nodal displacements. Once this has been done the matrix $[H]$ may be formed and the internal stresses evaluated. Since $[H]$ depends upon the dimensions of the element in question it is usually necessary to recalculate $[H]$ for each element.

Presentation of results

It is usual to print the input data both as a check and to enable a reader unfamiliar with the problem to interpret the results. In addition, the values of the nodal displacements and of the element stresses must be printed.

If the program is required for general use it is important that the output is in a clear and readable form. It is useless simply to present a designer with a mass of numbers! An adequate set of headings together with a clear presentation of the numerical results is therefore essential, a little time and thought spent at this stage being well rewarded later.

More advanced topics

Introduction

In the previous section the various stages involved in preparing a finite element computer program were explained. Of the eight stages, two are particularly important in that they either occupy the most computer time or decide the amount of computer storage necessary. It is desirable, both from the point of cost and from the point of minimising the delay between submitting the program and receiving the results, to economise as much as possible on these two factors. Therefore, some guidance on achieving this is now given. Fortunately, the two stages involved, namely the assembly of the $[K]$ matrix and the solution of the final equations, offer some scope in this direction and many different schemes are available to the programmer.

Assembly and solution

Due to the nature of the problem, provided that the nodes are numbered in a careful manner, the non-zero terms in $[K]$ will be concentrated in a narrow band situated adjacent to the leading diagonal. This fact, combined with the symmetrical nature of $[K]$ means that only a relatively small proportion of the matrix is of real interest. Provided advantage is taken of these factors, demands on the computer store may be considerably reduced. Moreover, if the solution procedure is so arranged that many of the operations involving the zero terms are eliminated, the speed of the solution can be increased. Methods which take advantage of the banded nature of $[K]$ are often referred to as 'banded methods.'

One such method is known as the tridiagonal method. In this the matrix is divided into a series of diagonal and off-diagonal blocks, as shown in Figure 10.6. Since these blocks contain all of the non-zero terms, it is necessary only to store these submatrices in the computer. Details of the solution procedure are well documented.[3] This method is fairly straightforward to program since the submatrices may be stored on magnetic tape and a standard matrix inversion routine used to perform this phase of the solution.

$$
\begin{Bmatrix} \{F_1\} \\ \{F_2\} \\ \{F_3\} \\ \vdots \\ \{F_k\} \\ \vdots \\ \{F_n\} \end{Bmatrix} =
\begin{bmatrix}
[K_1] & [C_1] & & & & \\
[C_1]^T & [K_2] & [C_2] & & & \\
& [C_2]^T & [K_2] & [C_3] & & \\
& & & \text{etc.,} & & \\
& & & & [C_{k-1}]^T [K_k] [C_k] & \\
& & & & \text{etc.,} & \\
& & & & & [C_{n-1}]^T [K_n]
\end{bmatrix}
\begin{Bmatrix} \{\delta_1\} \\ \{\delta_2\} \\ \{\delta_3\} \\ \vdots \\ \{\delta_k\} \\ \vdots \\ \{\delta_n\} \end{Bmatrix}
$$

Fig. 10.6. Layout of the $[K]$ matrix for the tridiagonal method

Methods which offer potentially greater economies are the so-called 'half-banded schemes'. In these the upper half of the diagonal band of the matrix is stored as a rectangular matrix as shown in Figure 10.7. The solution scheme, e.g. Cholesky's method[2], then proceeds to operate on this condensed form of the matrix. Such

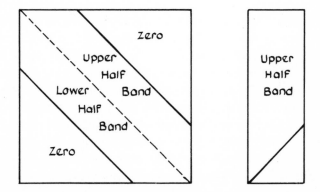

Fig. 10.7. Banded form of stiffness matrix

methods require careful programming to ensure that the correct terms in the matrix are used at all times.

One other type of method which may be used consists of that in which the assembly and solution proceed simultaneously.[4] An elimination technique is usually used, only that part of the matrix which is actually being operated upon being required at any one time.

Once the method of solution has been chosen, clearly the matrix must be assembled in the correct form. Thus the actual manner in which the assembly of $[K]$ is arranged is largely dictated by the method of solution to be adopted. For example, if the tridiagonal method is chosen then $[K]$ must be assembled in the form of the submatrix blocks $[K_i]$ and $[C_i]$.

The amount of computer storage required by each of the methods indicated above may be further reduced by the use of magnetic tape or disc to store large blocks of information not actually needed for the particular stage of the calculations currently in hand. For the tridiagonal method, since the solution procedure is such that it operates on a limited number of submatrices at any one time, it is convenient to store the submatrices on tape and to hold only those actually being operated on in the central store. As the calculations progress so the modified submatrices may be redeposited on tape and a new set of submatrices transferred to the central store.

Data preparation

As shown earlier, the amount of input data required by a finite element program is quite large and the preparation of this data therefore represents a formidable task. Indeed, for large problems involving several hundred elements this task may represent the greater part of the total time required to solve the problem. In cases where a number of similar problems are to be solved, it is useful to possess a routine which, given a small amount of data, generates the complete set of data for the finite element program. Such routines may even incorporate features enabling them to generate the element mesh automatically, even for very complex geometrical shapes.

As an alternative to actually producing data cards, the data preparation routine may be incorporated in the finite element program.

Plotting

For certain problems, where a large amount of output consisting of several hundred deflections and stresses is required, it may be desirable to allow the computer to produce these results in pictorial form by means of a graph plotter. In this way graphs of the principal stresses or contours of the deflections may be produced.

References

1. FADDEEV, D. K. and FADDEEVA, V. N. *Computational methods of linear algebra*. San Francisco, W. H. Freeman Co., 1963.
2. WEAVER, W. J. *Computer programs for structural analysis*. Princetown, N.J., D. Van Nostrand Co. Inc., 1967.
3. ROCKEY, K. C. and EVANS, H. R. A finite element solution for folded plate structures. International Conference on Space Structures, Battersea, London, 1966. Pp. 165–188 in *Space Structures*, edited by R. M. Davies. Oxford, Blackwell Scientific Publications, 1967.
4. BELLAMY, J. B. The analysis, design and optimisation of structural systems by a digital computer programming system. University of Wales PhD Thesis, 1969.

CHAPTER ELEVEN

Triangular Finite Element for Plate Flexure

Introduction

Triangular elements are more versatile than rectangular elements because they can be used for the analysis of plates having various boundary shapes, such as skew or curved bridge decks, and they also lend themselves more readily to mesh grading. In Chapter 7 the use of rectangular elements in plate flexure problems was discussed, and in the present chapter the use of triangular elements for the solution of these problems is considered. These triangular elements can then be combined with the triangular plane stress elements discussed in Chapter 5, in the same way as the rectangular elements were combined in Chapter 8 to give solutions for folded-plate and continuously curved shell structures.

The combination of triangular plane stress and plate flexure elements enables solutions to be obtained for folded plates and box girders of non-uniform cross-section, such as those shown in Figure 11.1. Such structures could not, of course, be analysed simply by using rectangular elements. Furthermore, rectangular elements can only be applied to the analysis of singly-curved shells, whereas triangular elements enable doubly-curved shells to be analysed also. The development of a triangular finite element for plate bending thus greatly enhances the scope of the finite element method.

Unfortunately, in attempting to obtain the stiffness matrix for a triangular plate bending element using the seven basic steps, certain difficulties are encountered. In fact, the first satisfactory solutions employing such an element were not obtained until several years after solutions using the other elements had been achieved. In this chapter the difficulties encountered in applying the standard procedure to the derivation of the stiffness matrix of a triangular plate bending element are first outlined.

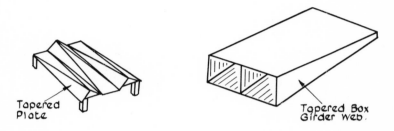

Fig. 11.1. Folded plates and box girders of non-uniform cross-section

Difficulties in deriving stiffness matrix

The co-ordinate and node numbering systems defined in Chapter 5 for a triangular element are again employed. These are as shown in Figure 11.2, the positions of the nodes then being defined by co-ordinates (x_1, y_1), (x_2, y_2) and (x_3, y_3).

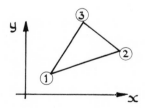

Fig. 11.2. Co-ordinate and node numbering system

From Chapter 7 it is also known that a plate bending element must have three degrees of freedom at each node, i.e. two rotations

(θ_x and θ_y) and a lateral displacement w. These are shown for node 1 of the triangular element in Figure 11.3, the positive directions being defined to correspond to those taken in Chapter 7. The triangular element thus has a total of nine degrees of freedom so that the element

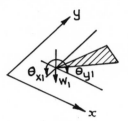

Fig. 11.3. Degrees of freedom at node 1

displacement and force vectors $\{\delta^e\}$ and $\{F^e\}$ each contain nine terms and the element stiffness matrix $[K^e]$ is a 9×9 matrix. Thus in choosing a function for the lateral displacement w, in accordance with the procedure previously established, the polynomial must include nine unknown constants, e.g.

$$w = \alpha_1 + \alpha_2 x + \alpha_3 y + \alpha_4 x^2 + \alpha_5 xy + \alpha_6 y^2 + \alpha_7 x^3 + \alpha_8 x^2 y + \alpha_9 y^3$$

Comparing this to a full cubic polynomial, it may be noted that one term has been omitted, namely the xy^2 term. The inclusion of the $x^2 y$ term rather than the xy^2 term is purely arbitrary so that, in fact, for no logical reason, the variation of w along a line on the element where x is constant differs from the variation along a line on the element where y is constant has been specified. This arrangement is obviously unsatisfactory as, for example, considering the particular case of the element shown in Figure 11.4 it would mean that the variation of w along edge 1–3 would differ from its variation along edge 1–2.

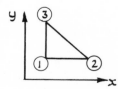

Fig. 11.4. Typical case

Several methods of overcoming this difficulty have been tried. For example, the ten terms of the full cubic expression can be retained in the expression for w, two of the coefficients being specified to be equal, e.g. $\alpha_8 = \alpha_9$. However, if this is done, the $[A]$ matrix in the expression $\{\delta^e\} = [A]\{\alpha\}$ becomes singular for certain orientations of the triangular element, e.g. when two sides of the triangle are parallel to the x and y axes. In such cases the $[A]$ matrix cannot be inverted and a solution cannot be obtained.

There is, in fact, no simple way around the problem and in order to obtain a solution the standard approach to the derivation of the element stiffness matrix, described in detail in earlier chapters, has to be changed, and an approach based on area co-ordinates used instead. This method is now outlined.

Derivation of element stiffness matrix using area co-ordinates

In order to define the term 'area co-ordinates', consider any point P having co-ordinates x and y taken inside a typical triangular element, as shown in Figure 11.5. When this point P is joined to the

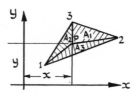

Fig. 11.5. Definition of area co-ordinates

three nodes of the element, the total area of the triangle is divided into three parts A_1, A_2, and A_3 as shown, A_1 being the area of the triangle directly opposite node 1, etc. The position of P is then uniquely defined by the area of the three small triangles. Denoting the total area of the triangular element by Δ, as in Chapter 5, the area co-ordinates (L_1, L_2 and L_3) of point P are defined as

$$L_1 = \frac{A_1}{\Delta} \qquad L_2 = \frac{A_2}{\Delta} \quad \text{and} \quad L_3 = \frac{A_3}{\Delta} \qquad (11.1)$$

Since $A_1+A_2+A_3=\Delta$ then $L_1+L_2+L_3=1$ so that once two of the area co-ordinates of a point are known the third is defined automatically.

Now, as in Chapter 5, it can be shown that the area of a triangle having nodal co-ordinates (x_1, y_1), (x_2, y_2) and (x_3, y_3), these being the co-ordinates of the complete triangular element, is given by

$$2\Delta = \det \begin{vmatrix} 1 & x_1 & y_1 \\ 1 & x_2 & y_2 \\ 1 & x_3 & y_3 \end{vmatrix}$$

That is,

$$\Delta = \tfrac{1}{2}[x_2 y_3 - y_2 x_3 + x_1(y_2 - y_3) + y_1(x_3 - x_2)]$$

A similar expression can be written for the smaller triangles. For example, for the triangle opposite to node 1, which has nodal co-ordinates (x, y), (x_2, y_2) and (x_3, y_3),

$$A_1 = \tfrac{1}{2}[x_2 y_3 - y_2 x_3 + x(y_2 - y_3) + y(x_3 - x_2)] \qquad (11.2)$$

Letting

$$a_1 = x_2 y_3 - y_2 x_3 \qquad b_1 = y_2 - y_3 \quad \text{and} \quad c_1 = x_3 - x_2 \qquad (11.3)$$

equation 11.2 may be simplified to give

$$A_1 = \tfrac{1}{2}(a_1 + b_1 x + c_1 y)$$

Substituting for A_1 in equation 11.1,

$$L_1 = \frac{a_1 + b_1 x + c_1 y}{2\Delta}$$

Similarly   (11.4)

$$L_2 = \frac{a_2 + b_2 x + c_2 y}{2\Delta} \quad \text{and} \quad L_3 = \frac{a_3 + b_3 x + c_3 y}{2\Delta}$$

where a_2, a_3, b_2, b_3, c_2 and c_3 can be obtained from equation 11.3 by changing the subscripts in cyclic order, i.e. $a_2 = x_3 y_1 - y_3 x_1$, $b_2 = y_3 - y_1$ and $c_2 = x_1 - x_3$, and so on. Equation 11.4 defines the area co-ordinates $(L_1, L_2$ and $L_3)$ of point P in terms of its Cartesian co-ordinates (x, y) and the nodal co-ordinates of the element (x_1, y_1), (x_2, y_2) and (x_3, y_3).

The next step is to choose a function to define the deflected shape of the element. This is an extremely important step as it is evident that the accuracy of a finite element solution depends on the extent to which the assumed deformation pattern is able to represent the actual deformation of the structure. Previously it has been shown that the accuracy of the finite element solution improves as the number of elements taken increases. This is true provided that a good deformation pattern has been chosen for the element as indeed has been the case with all the elements considered previously in this book. However, if a poor deformation pattern is chosen then, although the solutions converge towards a certain value as more and more elements are taken, this value may not be the correct value.

There are, in fact, two criteria[2] that can be of assistance in the choice of a displacement function. The function chosen should be such that it does not permit straining to occur within the element when the nodal displacements are caused by a rigid body movement. In addition, the function chosen should be able to express constant strain conditions within the element when the nodal displacements are compatible with a constant strain condition.

The second of these two conditions is the more rigorous. In fact, it incorporates the first, since a rigid body displacement is a particular case of a constant strain condition in which the constant strain has a zero value.

Provided that these two conditions are satisfied by the chosen displacement function, a finite element solution will converge towards the correct result as the number of elements taken is increased. This is true for all finite element solutions, not only for the triangular plate bending solution now being considered. The two convergence criteria now have to be considered explicitly because a suitable displacement function is not immediately apparent. However, it should be noted that all the functions chosen in Chapters 5, 6 and 7 do, in fact, satisfy the required criteria.

In the present case, a convenient way of satisfying the first of these conditions is to consider the displacement of the element to be made up of two parts, as shown in Figure 11.6. In the first part, the element is considered to undergo a pure displacement as a rigid body, no curvatures (i.e. strains) being set up within the element. A suitable function is then chosen to represent this movement. In the second part, the element is considered to be simply supported at the nodes,

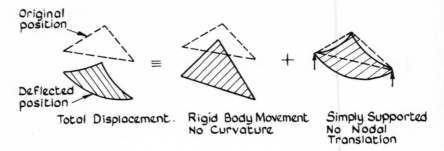

Fig. 11.6. Deformations of typical element

so that no lateral displacement of the nodes and consequently no rigid body movement can occur. Another function is then chosen to define this behaviour. The complete displacement of the element is then obtained by superimposing the two parts.

Since no curvatures are permitted to occur in the first part where the rigid body movements occur, and no rigid body movements are permitted in the second part where the curvatures take place, self-straining due to rigid body movements is eliminated in accordance with the requirements of the first condition.

Denoting the lateral displacement at any point on the element during the rigid body movement by w^{rb} and that during the simply supported case as w^{ss}, the total lateral displacement can be expressed as

$$w = w^{rb} + w^{ss} \tag{11.5}$$

Considering the rigid body movement by itself, this is of the form shown in Figure 11.7.

Since no curvatures are set up, the lateral displacement must obviously be a linear function of x and y, and since curvature is equal

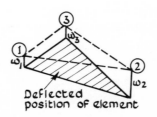

Fig. 11.7. Rigid body movement of element

to the second differential of the displacement it is then zero. From Figure 11.7 it is also obvious that the lateral displacement at any point on the element is governed by the nodal displacements w_1, w_2 and w_3 of the element. Since the area co-ordinates defined in equation 11.4 are linear functions of x and y, the lateral displacement at any point may be simply defined in terms of these area co-ordinates and the nodal displacements as

$$w^{rb} = w_1 L_1 + w_2 L_2 + w_3 L_3 \tag{11.6}$$

To check whether this function is suitable, consider the case when the general point P coincides with node 1. In this case, by definition, $L_2 = L_3 = 0$, $L_1 = 1$ and then $w^{rb} = w_1$ as required.

Considering the simply supported case by itself, the element now has only two rotational degrees of freedom at each node, i.e. θ_x and θ_y, since no lateral displacements are allowed. The element displacement vector then only contains six terms and may be written as

$$\{\delta^{ess}\} = \begin{Bmatrix} \theta_{x1}^{ss} \\ \theta_{y1}^{ss} \\ \hline \theta_{x2}^{ss} \\ \theta_{y2}^{ss} \\ \hline \theta_{x3}^{ss} \\ \theta_{y3}^{ss} \end{Bmatrix} \tag{11.7}$$

The corresponding force vector $\{F^{ess}\}$ also contains only six terms and the stiffness matrix for the simply supported element $[K^{ess}]$ is then a 6×6 matrix. Now as is known from Step 2 in Chapter 7, the slopes at any point can be related to the displacements at that point as follows.

$$\theta_x^{ss} = -\frac{\partial w^{ss}}{\partial y} \quad \text{and} \quad \theta_y^{ss} = \frac{\partial w^{ss}}{\partial x}$$

Substituting for w^{ss} from equation 11.5,

$$\theta_x^{ss} = -\frac{\partial}{\partial y}(w - w^{rb}) = -\frac{\partial w}{\partial y} + \frac{\partial w^{rb}}{\partial y} = \theta_x + \frac{\partial w^{rb}}{\partial y} \qquad (11.8)$$

Similarly

$$\theta_y^{ss} = \theta_y - \frac{\partial w^{rb}}{\partial x}$$

where θ_x and θ_y are the slopes at the point under consideration when the total displacement w occurs in the element.

Substituting the expression assumed for w^{rb} in equation 11.6 into equation 11.8 the slopes at a point in the simply supported case can be expressed in terms of the total slopes at that point and the total lateral displacements as follows.

$$\theta_x^{ss} = \theta_x + \frac{1}{2\Delta}(w_1 c_1 + w_2 c_2 + w_3 c_3)$$

and $\qquad\qquad\qquad\qquad\qquad\qquad\qquad\qquad\qquad\qquad\qquad\qquad$ (11.9)

$$\theta_y^{ss} = \theta_y - \frac{1}{2\Delta}(w_1 b_1 + w_2 b_2 + w_3 b_3)$$

Having determined the slopes at any point, the nodal rotations in the simply supported case can then be expressed simply in terms of the total nodal displacements. For example, at node 1,

$$\theta_{x1}^{ss} = \theta_{x1} + \frac{1}{2\Delta}(w_1 c_1 + w_2 c_2 + w_3 c_3)$$

and

$$\theta_{y1}^{ss} = \theta_{y1} - \frac{1}{2\Delta}(w_1 b_1 + w_2 b_2 + w_3 b_3)$$

Similar expressions can be obtained for the other two nodes, and the complete relationship between the nodal rotations of the simply supported element and those of the complete element may be written in matrix form as

$$
\{\delta^{ess}\} =
\begin{Bmatrix}
\theta^{ss}_{x1} \\
\theta^{ss}_{y1} \\
\theta^{ss}_{x2} \\
\theta^{ss}_{y2} \\
\theta^{ss}_{x3} \\
\theta^{ss}_{y3}
\end{Bmatrix}
= \frac{1}{2\Delta}
\begin{bmatrix}
2\Delta & 0 & c_1 & 0 & 0 & c_2 & 0 & 0 & c_3 \\
0 & 2\Delta & -b_1 & 0 & 0 & -b_2 & 0 & 0 & -b_3 \\
0 & 0 & c_1 & 2\Delta & 0 & c_2 & 0 & 0 & c_3 \\
0 & 0 & -b_1 & 0 & 2\Delta & -b_2 & 0 & 0 & -b_3 \\
0 & 0 & c_1 & 0 & 0 & c_2 & 2\Delta & 0 & c_3 \\
0 & 0 & -b_1 & 0 & 0 & -b_2 & 0 & 2\Delta & -b_3
\end{bmatrix}
\begin{Bmatrix}
\theta_{x1} \\
\theta_{y1} \\
w_1 \\
\theta_{x2} \\
\theta_{y2} \\
w_2 \\
\theta_{x3} \\
\theta_{y3} \\
w_3
\end{Bmatrix}
$$

$$(11.10)$$

This can be summarised as

$$\{\delta^{ess}\} = [T]\{\delta^e\} \tag{11.11}$$

A relationship may also be obtained between the nodal forces in the simply supported case and the nodal forces of the complete element by the application of the principle of virtual work. Since no curvatures are set up during the rigid body displacement, no internal work is done during this part of the total displacement. Thus during any virtual displacement the work done in the simply supported case is equal to the total work done. Therefore

$$\{\delta^{ess}\}^T\{F^{ess}\} = \{\delta^e\}^T\{F^e\}$$

Substituting for $\{\delta^{ess}\}$ from equation 11.11,

$$[T]^T\{\delta^e\}^T\{F^{ess}\} = \{\delta^e\}^T\{F^e\}$$

Therefore

$$[T]^T\{F^{ess}\} = \{F^e\} \tag{11.12}$$

Now, as discussed earlier, the element nodal displacements and forces in the simply supported case are related by a 6×6 stiffness matrix $[K^{ess}]$, i.e.

$$\{F^{ess}\} = [K^{ess}]\{\delta^{ess}\} \tag{11.13}$$

Substituting for $\{F^{ess}\}$ from equation 11.13 in equation 11.12 and then substituting for $\{\delta^{ess}\}$ from equation 11.11, the following relationship is obtained.

$$\{F^e\} = [T]^T[K^{ess}][T]\{\delta^e\} \tag{11.14}$$

Also, for the complete element, it is known that $\{F^e\} = [K^e]\{\delta^e\}$, so that the real stiffness matrix $[K^e]$ that is required for the triangular element is related to the stiffness matrix for the simply supported element as follows.

$$[K^e] = [T]^T[K^{ess}][T] \tag{11.15}$$

$[T]$ is already known from equation 11.10. Hence, before the complete stiffness matrix for the element $[K^e]$ can be determined, the stiffness matrix $[K^{ess}]$ for the simply supported element must be derived. The steps in this derivation are similar in principle to the seven basic steps shown in Appendix 1, although they now differ in many details since area co-ordinates are being used.

Derivation of stiffness matrix for simply supported element

First of all, an expression to define the lateral displacement at a point on the simply supported element in terms of the nodal displacement is chosen as follows.

$$w^{ss} = N_{x1}\,\theta^{ss}_{x1} + N_{y1}\,\theta^{ss}_{y1} + N_{x2}\,\theta^{ss}_{x2} + N_{y2}\,\theta^{ss}_{y2} + N_{x3}\,\theta^{ss}_{x3} + N_{y3}\,\theta^{ss}_{y3}$$

$$\tag{11.16}$$

Each individual N_x and N_y term in this expression in fact represents a function and these are known as 'shape functions'. Since these shape functions directly relate the displacements at any point on the element to the nodal displacements of the element, they correspond to the product $[f(x, y)][A]^{-1}$ in step III, equation III of the basic procedure described in the earlier chapters.

The slopes θ^{ss}_x and θ^{ss}_y at this point are then obtained from $\theta^{ss}_x = -\partial w^{ss}/\partial y$ and $\theta^{ss}_y = \partial w^{ss}/\partial x$, that is

$$\theta^{ss}_x = -\left[\frac{\partial N_{x1}}{\partial y}\,\theta^{ss}_{x1} + \frac{\partial N_{y1}}{\partial y}\,\theta^{ss}_{y1} + \frac{\partial N_{x2}}{\partial y}\,\theta^{ss}_{x2}\right.$$

$$\left. + \frac{\partial N_{y2}}{\partial y}\,\theta^{ss}_{y2} + \frac{\partial N_{x3}}{\partial y}\,\theta^{ss}_{x3} + \frac{\partial N_{y3}}{\partial y}\,\theta^{ss}_{y3}\right]$$

and

$$\theta_y^{ss} = \frac{\partial N_{x1}}{\partial x}\theta_{x1}^{ss} + \frac{\partial N_{y1}}{\partial x}\theta_{y1}^{ss} + \frac{\partial N_{x2}}{\partial x}\theta_{x2}^{ss} + \frac{\partial N_{y2}}{\partial x}\theta_{y2}^{ss}$$

$$+ \frac{\partial N_{x3}}{\partial x}\theta_{x3}^{ss} + \frac{\partial N_{y3}}{\partial x}\theta_{y3}^{ss}$$

The individual shape functions must be chosen so that they satisfy boundary conditions at the nodes of the simply supported element. For example, at node 1 these conditions are $w^{ss}=0$, $\theta_x^{ss}=\theta_{x1}^{ss}$ and $\theta_y^{ss}=\theta_{y1}^{ss}$. Summarising all these conditions for all the nodes it is found that, for example, the function N_{x1} has to satisfy the following requirements.

$$\left.\begin{array}{l} N_{x1} = 0 \text{ at all nodes} \\[2mm] \dfrac{\partial N_{x1}}{\partial y} = \begin{array}{l} -1 \text{ at node 1} \\ 0 \text{ at nodes 2 and 3} \end{array} \\[4mm] \dfrac{\partial N_{x1}}{\partial x} = 0 \text{ at all nodes} \end{array}\right\} \qquad (11.17)$$

The conditions for the other functions are similar.

The simplest functions that satisfy these boundary conditions are

$$\left.\begin{array}{l} N_{x1} = b_3 L_1{}^2 L_2 - b_2 L_1{}^2 L_3 \\[3mm] N_{y1} = c_3 L_1{}^2 L_2 - c_2 L_1{}^2 L_3 \end{array}\right\} \qquad (11.18)$$

and

where the b and c terms are as defined in equation 11.3. Corresponding functions for N_{x2}, N_{y2}, N_{x3} and N_{y3} can be obtained from equation 11.18 by changing the subscripts in cyclic order, e.g. $N_{x2}=b_1 L_2{}^2 L_3 -b_3 L_2{}^2 L_1$, etc. These functions can be proved to satisfy the required boundary conditions by a simple substitution into equations similar to equation 11.17. For example, the function chosen for N_{x1} is now checked to see whether it satisfies the necessary conditions.

From the definition of the area co-ordinates in equation 11.1, it is seen that two of these are zero at any node, e.g. at node 1, $L_2=L_3=0$.

Therefore, from equation 11.18, $N_{x1}=0$ at each node, thus satisfying the first condition of equation 11.17. Also,

$$\frac{\partial N_{x1}}{\partial y} = \frac{\partial}{\partial y}(b_3 L_1{}^2 L_2 - b_2 L_1{}^2 L_3)$$

$$= \frac{\partial}{\partial y}\left[\frac{b_3(a_2+b_2 x+c_2 y)-b_2(a_3+b_3 x+c_3 y)}{2\Delta}\right]L_1{}^2$$

$$= \frac{1}{2\Delta}[(b_3 c_2 - b_2 c_3)L_1{}^2 + (b_3 L_2 - b_2 L_3)2L_1 c_1]$$

Now

$$(b_3 c_2 - b_2 c_3)L_1{}^2 = \begin{cases} 0 \text{ at nodes 2 and 3, since } L_1=0 \\ (b_3 c_2 - b_2 c_3) \text{ at node 1, since } L_1=1 \end{cases}$$

$$(b_3 L_2 - b_2 L_3)2L_1 c_1 = \begin{cases} 0 \text{ at all nodes, since both values of } L \text{ are} \\ \text{zero, at each node} \end{cases}$$

Therefore $\partial N_{x1}/\partial y = 0$ at nodes 2 and 3, and

$$\frac{\partial N_{x1}}{\partial y} = \frac{1}{2\Delta}[b_3 c_2 - b_2 c_3] =$$
$$\frac{1}{2\Delta}[(y_1 - y_2)(x_1 - x_3) - (y_3 - y_1)(x_2 - x_1)] = -1$$

Hence, the second condition of equation 11.17 is satisfied by the function chosen for N_{x1} and it can be proved in a similar way that this function also satisfies the last necessary condition.

These functions also ensure that the lateral displacements are continuous along element interfaces although, in common with the displacement function chosen for the rectangular plate bending element, they do not preserve continuity of normal slopes.

Unfortunately, the chosen functions have been proved to be incapable of satisfying the constant strain criterion which, as discussed earlier, is necessary for convergence. However, if the functions given in equation 11.18 are modified as follows,

$$\left.\begin{aligned} N_{x1} &= b_3(L_1{}^2 L_2 + \tfrac{1}{2}L_1 L_2 L_3) - b_2(L_1{}^2 L_3 + \tfrac{1}{2}L_1 L_2 L_3) \\ N_{y1} &= c_3(L_1{}^2 L_2 + \tfrac{1}{2}L_1 L_2 L_3) - c_2(L_1{}^2 L_3 + \tfrac{1}{2}L_1 L_2 L_3) \end{aligned}\right\} \quad (11.19)$$

it is found that the constant strain criterion is now satisfied. N_{x2}, N_{y2}, N_{x3} and N_{y3} are again obtained from equation 11.19 by changing the subscripts in cyclic order. It should be noted that the functions of equation 11.19 differ from those of equation 11.18 only by the addition of terms which are functions of the product $L_1 L_2 L_3$. The function $L_1 L_2 L_3$ gives zero values of deflections and slopes at the nodes, so that its addition does not violate any of the boundary condition requirements.

The functions given in equation 11.19 are thus satisfactory and on substitution into equation 11.16 the lateral displacement and consequently the slopes at any point in the element are defined in terms of the nodal displacements.

The state of displacement at any point in the element has thus been defined in terms of the nodal displacements so a stage corresponding to step III in the basic procedure described in earlier chapters has now been reached. The curvatures set up in the element may now be related to nodal displacements according to equation 7.6 in step IV of Chapter 7, i.e.

$$\{\varepsilon^{ss}(x, y)\} = \begin{Bmatrix} -\partial^2 w^{ss}/\partial x^2 \\ -\partial^2 w^{ss}/\partial y^2 \\ 2\partial^2 w^{ss}/\partial x\, \partial y \end{Bmatrix}$$

where the suffix 'ss' indicates that the simply supported element is still under consideration. This leads to general equation IV, i.e.

$$\{\varepsilon^{ss}(x, y)\} = [B^{ss}]\{\delta^{ess}\}$$

where the matrix $[B^{ss}]$ is now a 3×6 matrix, and can be determined by substituting the expression for w^{ss} from equation 11.16. For ease of presentation, the $[B^{ss}]$ matrix thus obtained may be written in a partitioned form.

$$[B^{ss}] = [[B_1{}^{ss}]\,[B_2{}^{ss}]\,[B_3{}^{ss}]]$$

where

$$[B_1{}^{ss}] = \begin{bmatrix} -\dfrac{\partial^2 N_{x1}}{\partial x^2} & -\dfrac{\partial^2 N_{y1}}{\partial x^2} \\[2ex] -\dfrac{\partial^2 N_{x1}}{\partial y^2} & -\dfrac{\partial^2 N_{y1}}{\partial y^2} \\[2ex] \dfrac{2\partial^2 N_{x1}}{\partial x\, \partial y} & \dfrac{2\partial^2 N_{y1}}{\partial x\, \partial y} \end{bmatrix} \qquad (11.20)$$

the values of shape functions N_{x1} and N_{y1} being defined in equation 11.19. The two other partitions have a similar form to that shown for $[B_1{}^{ss}]$, but $[B_2{}^{ss}]$ involves terms in N_{x2} and N_{y2} and $[B_3{}^{ss}]$ involves terms in N_{x3} and N_{y3}.

Next, as in basic step V the internal moments set up in the element are related to the curvatures, the $[D]$ matrix being as defined in equation 7.12.

Finally, as in step VI, the stiffness matrix for the simply supported element is obtained from general equation VI as

$$[K^{ess}] = \int [B^{ss}]^T [D][B^{ss}] \, d(\text{vol}) \qquad (11.21)$$

which for the case of plate bending, may be written as

$$[K^{ess}] = \iint [B^{ss}]^T [D][B^{ss}] \, dx \, dy$$

The $[K^{ess}]$ matrix obtained will be a 6×6 matrix because the simply supported element has six degrees of freedom.

Having thus found the stiffness matrix for the simply supported element, the stiffness matrix for the actual element $[K^e]$ can be obtained from equation 11.15. $[K^e]$ will be a 9×9 matrix since the actual element has a total of nine degrees of freedom.

Also, since the $[D]$ and $[B^{ss}]$ matrices are known, the stress–displacement matrix for the simply supported element can be obtained as in general step VII as $[H^{ss}] = [D][B^{ss}]$. Since no internal curvatures are set up during the rigid body displacement then no internal moments are set up in this case either. Consequently, the moments calculated for the simply supported element using the $[H^{ss}]$ matrix are the internal moments required for the actual element.

The final value obtained for the element stiffness matrix $[K^e]$ is not presented because the terms of the matrix are long and complex expressions. Since the operations involved in setting up these terms are often of a cyclic nature, it is usual to use the computer to carry out some of these operations.

Summary of derivation procedure

Because the explicit form of the $[K^e]$ matrix is not presented, the steps that must be followed to derive it from the information given in this chapter are now summarised.

Step I: Take values of a_1, b_1, c_1, etc. from equation 11.3 and sub-stitute in equation 11.4 to give the area co-ordinate values L_1, L_2 and L_3.

Step II: Substitute for L_1, L_2 and L_3 and for a_1, b_1, c_1, etc. in equation 11.19 to give the values of the shape functions N_{x1}, N_{y1}, etc.

Step III: Substitute for N_{x1}, N_{y1}, etc. in equation 11.20 and carry out the required differentiation, thus establishing the $[B^{ss}]$ matrix.

Step IV: Substitute for $[B^{ss}]$ in equation 11.21 and evaluate the product $[B^{ss}]^T [D][B^{ss}]$.

Step V: Integrate this product over the area of the element to obtain the stiffness matrix for the simply supported element $[K^{ess}]$.

Step VI: Evaluate the $[T]$ matrix as in equation 11.10.

Step VII: Evaluate the required stiffness matrix $[K^e]$ for the element according to equation 11.15, i.e. $[K^e]=[T]^T[K^{ess}][T]$.

Step VIII: Evaluate the stress–displacement matrix $[H]$ for the element $[H]=[H^{ss}]=[D][B^{ss}]$.

Conclusions

It is apparent that the derivation of a stiffness matrix for a tri-angular element in plate bending is considerably more complex than for the elements considered earlier in this book. However, once the stiffness matrix for the triangular element has been established, the remaining steps in obtaining a complete finite element solution using this element are the same as for any other element.

References

1. BAZELEY, G. P., CHEUNG, Y. K., IRONS, B. M. and ZIENKIEWICZ, O. C. Triangular elements in plate bending—conforming and non-conforming solutions. *Proceedings of the first conference on*

Matrix Methods in Structural Mechanics at Wright-Patterson Air Force Base, Ohio, 1965.
2. IRONS, B. M. and DRAPER, K. J. Inadequacy of nodal connections in a stiffness solution for plate bending. *AIAA Journal*, Vol. 3, No. 5, 1965. p. 961.

CHAPTER TWELVE

Additional Topics

Introduction

The elements developed in the previous chapters are the basic elements ranging from a simple truss element to a plane shell element. These elements have been applied to the analysis of a wide range of structures and have been found to give accurate solutions. However, for a variety of reasons, many other elements have been developed. The type of problem or the shape of the structure may make it necessary or desirable to develop special elements, or to combine different elements together. For example, for the analysis of arbitrary shaped solids, three-dimensional elements are required. Starting with the tetrahedron and the brick element, developments have been made with curved and irregularly-shaped tetrahedra and hexahedra, e.g. the isoparametric elements.

In addition to the development of elements for application to special types of problems, work has been done on the development of 'higher-order' finite elements, the object being to achieve more accurate solutions with fewer elements. At the time of writing, there is some debate about the merits of introducing the greater computational complexity involved in using the higher-order elements as opposed to the employment of a greater number of simpler elements.

This chapter contains a discussion of some of the higher-order finite elements and three-dimensional elements. In addition, some brief comments are made about a mixed formulation. Readers can obtain more detailed information from some of the references in the Bibliography.

Higher-order elements

It was stated in Chapter 4 that individual finite elements are constrained to deform in a specified pattern, given by the displacement function. In Chapters 5, 6 and 7, examples have been given of the displacement functions used for triangular and rectangular elements. It was shown that, in most cases, continuity of displacement was maintained along the boundaries of adjacent elements by specifying the displacement parameters at the corner nodes. For example, with the plane elasticity triangle, the in-plane displacements u and v in the co-ordinate directions x and y were specified at each node, giving the triangular element six degrees of freedom. This means that the displacement functions have to be specified with only six undetermined coefficients $\{\alpha\}$ and linear polynomials in x and y are selected for both u and v (equation 5.3). It is seen that the resulting element is a constant strain element. This has certain disadvantages, particularly in regions of high stress gradient where the attempt is being made to use a series of 'steps' to approximate a steep curve. One method of dealing with this problem is to use a very fine mesh while still using the basic element. However an alternative solution is to use 'higher-order' elements, i.e. elements having higher-order polynomials for the displacement functions. This can be done by specifying additional displacement parameters for the element, thus giving it more degrees of freedom and thereby enabling higher-order polynomials to be selected for the displacement functions. There are two approaches commonly used. Firstly, additional displacement parameters are specified at the corner nodes. In the case of the plane elasticity triangular element, for example, some or all of the displacement derivatives can be specified, so that at each node there can be up to six parameters, i.e. u, v, $\partial u/\partial x$, $\partial v/\partial y$, $\partial u/\partial y$ and $\partial v/\partial x$. The three-node element now has up to eighteen degrees of freedom, permitting a higher-order polynomial for the displacement function.

Element	No. of Nodes	Nodal Displacements	Element Degrees of Freedom	Displacement Function
	3	u, v	6	Linear
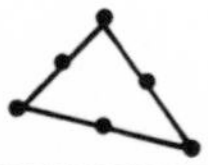	6	u, v	12	Parabolic
	9	u, v	18	Cubic
	3	$u, \dfrac{\partial u}{\partial x}, \dfrac{\partial u}{\partial y}$ $v, \dfrac{\partial v}{\partial x}, \dfrac{\partial v}{\partial y}$	18	Cubic

Fig. 12.1. Some plane elasticity triangular element types

Element	No. of Nodes	Nodal Displacements	Element Degrees of Freedom.	Degree of Displacement Function Polynominal.
	3	$w, \dfrac{\partial w}{\partial x}, \dfrac{\partial w}{\partial y}$	9	3
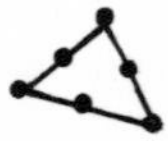	6	corner: $w, \dfrac{\partial w}{\partial x}, \dfrac{\partial w}{\partial y}$ midside: $w, \dfrac{\partial w}{\partial n}$	15	4
	3	$w, \dfrac{\partial w}{\partial x}, \dfrac{\partial w}{\partial y}$ $\dfrac{\partial^2 w}{\partial x^2}, \dfrac{\partial^2 w}{\partial x \partial y}, \dfrac{\partial^2 w}{\partial y^2}$	18	5

Fig. 12.2. Some plate bending triangular element types

Secondly, in addition to the element corner nodes, nodes may be placed along the sides or inside the element, and at these nodes any desired displacement parameters can be specified, thus increasing the number of degrees of freedom of the element.

Figures 12.1 and 12.2 illustrate some of the types of elements which have been used for plane elasticity and plate bending respectively. Further details may be found in references 1 to 4 and in the Bibliography.

These two approaches, i.e. allowing more degrees of freedom at the corner node or using additional side or internal nodes, have been used for various element shapes in both two and three dimensions. These elements have the advantage that fewer of them are required to idealise the structure. However, this is achieved at the expense of somewhat greater computational complexity.

Three-dimensional elements

The application of the finite element method in the three-dimensional theory of elasticity is a straightforward extension of the techniques described in Chapters 5 and 6 for plane elasticity problems. The special case of a body of revolution has already been considered in Chapter 9, where, owing to the axial symmetry, the three-dimensional problem becomes effectively a two-dimensional problem.

For three-dimensional problems, the nodal points are specified in position by three Cartesian co-ordinates (x, y, z). At least three nodal parameters, u, v and w for displacements, and the forces F_x, F_y and F_z must be specified, as in equations 12.1 and 12.2.

$$\{\delta\} = \begin{Bmatrix} u \\ v \\ w \end{Bmatrix} \tag{12.1}$$

$$\{F\} = \begin{Bmatrix} F_x \\ F_y \\ F_z \end{Bmatrix} \tag{12.2}$$

The most elementary solid element is the tetrahedron, having four nodes, as shown in Figure 12.3. This element has twelve degrees of

Fig. 12.3. Tetrahedral element

freedom when three displacement parameters are specified. A complete first-degree polynomial in x, y and z is required for the displacement function (step II) (see equations 12.3).

$$\left.\begin{aligned}
u &= \alpha_1 + \alpha_2 x + \alpha_3 y + \alpha_4 z \\
v &= \alpha_5 + \alpha_6 x + \alpha_7 y + \alpha_8 z \\
w &= \alpha_9 + \alpha_{10} x + \alpha_{11} y + \alpha_{12} z
\end{aligned}\right\} \tag{12.3}$$

Substituting for the nodal displacements enables the coefficients $\{\alpha\}$ to be determined (step III). From the theory of elasticity the strain–displacement relationship, as required in step IV, for three dimensions is

$$\{\varepsilon(x, y, z)\} = \left\{\begin{array}{c}
\varepsilon_x \\
\varepsilon_y \\
\varepsilon_z \\
\gamma_{xy} \\
\gamma_{yz} \\
\gamma_{zx}
\end{array}\right\} = \left\{\begin{array}{c}
\dfrac{\partial u}{\partial x} \\
\dfrac{\partial v}{\partial y} \\
\dfrac{\partial w}{\partial z} \\
\dfrac{\partial u}{\partial y} + \dfrac{\partial v}{\partial x} \\
\dfrac{\partial v}{\partial z} + \dfrac{\partial w}{\partial y} \\
\dfrac{\partial w}{\partial x} + \dfrac{\partial u}{\partial z}
\end{array}\right\} \tag{12.4}$$

and the stress strain $[D]$ matrix as required in step V is

$$[D] = \frac{E(1-v)}{(1+v)(1-2v)} \begin{bmatrix} 1 & \dfrac{v}{1-v} & \dfrac{v}{1-v} & 0 & 0 & 0 \\ & 1 & \dfrac{v}{1-v} & 0 & 0 & 0 \\ & & 1 & 0 & 0 & 0 \\ & & & \dfrac{1-2v}{2(1-v)} & 0 & 0 \\ \text{symmetric} & & & & \dfrac{1-2v}{2(1-v)} & 0 \\ & & & & & \dfrac{1-2v}{2(1-v)} \end{bmatrix}$$

$$(12.5)$$

The $[B]$ matrix is obtained as in the two-dimensional elasticity cases, and the element stiffness matrix is again given in step VI by

$$[K^e] = \int [B]^T [D] [B] \, d(\text{vol})$$

It is seen that for the tetrahedron element the strain is constant within the element thus simplifying the determination of $[K^e]$. The algebra is, however, more laborious than for the two-dimensional case, as are also the structural idealisation and the preparation of input data.

After the tetrahedron with four nodes, the next development is the brick element, which has eight nodes as shown in Figure 12.4. The basic element now has twenty-four degrees of freedom, permitting a higher-order polynomial to be used for the displacement function. However this element has only limited application since it does not fit irregular boundaries and it is difficult to grade the mesh.

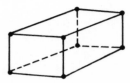

Fig. 12.4. Brick element

A major development in three-dimensional elements occurred with the introduction of curvilinear co-ordinates. These enable the simple tetrahedra and hexahedra to be distorted into arbitrary shapes that are better suited to fitting the types of boundaries which occur in practice (Figure 12.5).

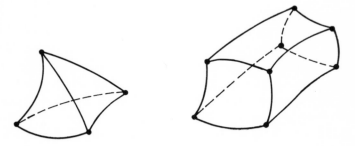

Fig. 12.5. Element defined in curvilinear co-ordinates

The basic elements in the Cartesian co-ordinate system (x, y, z) can be mapped in a new curvilinear system (ξ, η, ζ) provided suitable co-ordinate transformation relationships are known. One method of establishing the co-ordinate transformations is to use the so-called 'shape function' of the element. The shape function, as already described in Chapter 11, relates the displacement at any point to the nodal displacements, and is given in step III, equation III by $[f(x, y)][A]^{-1}$. This method has led to the development of the isoparametric elements,[1,2] which have been used to analyse three-dimensional problems such as dams and thick shell structures. In the development of the element stiffness matrices, numerical integration is now essential owing to the algebraic complexity.

Mixed formulations

In the displacement (or stiffness) method of analysis which has been used throughout this book, the displacements are taken as the unknown quantities, geometrically compatible states in the individual elements being combined to give a system of equilibrium equations. Another method is the force (or flexibility) method, in which the forces are treated as the unknown quantities, equilibrium states in the individual elements being combined to give a system of geometrical compatibility equations. These two 'pure' methods are the most familiar in finite element analysis, the displacement method having gained in prominence.

At an early stage in its development it was realised that the finite element displacement method represented an application of the well-known variational principle of minimum potential energy. The relationship between finite element methods and the well-established classical methods of structural analysis has encouraged researchers to seek to use variational methods to develop further the finite element method. The force method is an application of the variational principle of minimum complementary potential energy.

So-called mixed formulations, in which simultaneous use is made of both displacement and force variables and resulting in mixed form system equations, have been adopted by some researchers. This method allows the analyst to make simultaneous assumptions on both displacements and forces (stresses).[1]

Conclusions

This text has been planned to form an elementary introduction to the use of the finite element method. The Bibliography may help to guide readers to further information on the method. There is an extensive literature in the technical press of research papers on many aspects of the development and application of the finite element method. The Bibliography has been limited to the main sources, the text books and conference proceedings that are the basis for further reading.

It is perhaps fitting finally to mention some of the areas of engineering where the finite element method has been successfully applied. These include the stress and displacement analysis of engineering structures and machine components, including plasticity, creep and large displacement behaviour; stability, vibrations, shell and folded plate problems; stresses in soils and rocks; heat transfer, fluid flow and seepage flow problems. Such methods are now also being applied to problems of magnetism and electricity. The finite element method is therefore a very powerful tool for the analysis of a wide range of engineering problems.

References

1. HOLAND, I. and BELL, K. *Finite element methods in stress analysis.* Trondheim, Norway, Tapir, 1969.
2. IRONS, B. M. and ZIENKIEWICZ, O. C. The isoparametric element system—a new concept in finite element analysis. Conference on Recent Advances in Stress Analysis. London, British Constructional Steelwork Association/Royal Aeronautical Society, March 1968.
3. CLOUGH, R. W. The finite element method in structural mechanics. In *Stress analysis*, edited by O. C. Zienkiewicz and G. S. Holister. London, John Wiley & Sons Ltd, 1965.
4. ZIENKIEWICZ, O. C. *The finite element method in engineering science.* London, McGraw-Hill Publishing Co. Ltd, 1971.

Bibliography

The following books and conference proceedings provide good reference data for further reading.

AMERICAN SOCIETY OF CIVIL ENGINEERS. *Second conference on electronic computation*. Pittsburgh, September 1960.

ARGYRIS, J. H. *Recent advances in matrix methods of structural analysis*. London, Pergamon Press, 1964.

FRAEIJS DE VEUBEKE, B. (Editor). *Matrix methods of structural analysis*. AGARD. London, Pergamon Press, 1964.

GALLAGHER, R. H. *A correlation study of methods of matrix structural analysis*. AGARD. London, Pergamon Press, 1964.

ZIENKIEWICZ, O. C. and HOLISTER, G. S. (Editors). *Stress analysis, recent developments in numerical and experimental methods*. London, John Wiley & Sons Ltd, 1965.

WRIGHT–PATTERSON AIR FORCE BASE. *Proceedings of conferences on matrix methods in structural mechanics*. First conference 1965. Second conference 1968. Third conference 1971.

HOLAND, I. and BELL, K. (Editors). *Finite element methods in stress analysis*. Trondheim, Norway, Tapir, 1969.

AMERICAN SOCIETY OF CIVIL ENGINEERS. *Applications of finite element method in civil engineering*. Vanderbilt University conference, 1969.

TOTTENHAM, H. and BREBBIA, C. (Editors). *Finite element techniques in structural mechanics*. Southampton University, 1970.

ZIENKIEWICZ, O. C. *The finite element method in engineering science*. Maidenhead, McGraw-Hill Publishing Co. Ltd, 1971.

DESAI, C. S. and ABEL, J. F. *Introduction to the finite element method*. London, Van Nostrand Reinhold Co. Ltd, 1972.

Basic Steps in the Derivation of the Element Stiffness Characteristics

Step I

Identify the type of problem.
Choose a suitable co-ordinate system.
Number the nodes.

$$\{F^e\} = [K^e]\{\delta^e\} \tag{I}$$

Step II

Choose a displacement function.

$$\{\delta(x, y)\} = [f(x, y)]\{\alpha\} \tag{II}$$

Step III

Obtain the state of displacement at any point in terms of the nodal displacements.

$$\{\delta(x, y)\} = [f(x, y)][A]^{-1}\{\delta^e\} \tag{III}$$

Step IV

Relate the strains at any point to $\{\delta(x, y)\}$ and hence to $\{\delta^e\}$

$$\{\varepsilon(x, y)\} = [B]\{\delta^e\} \tag{IV}$$

Step V

Relate the stresses at any point to $\{\varepsilon(x, y)\}$ and hence to $\{\delta^e\}$

$$\{\sigma(x, y)\} = [D][B]\{\delta^e\} \tag{V}$$

Step VI

Replace $\{\sigma(x, y)\}$ by equivalent nodal loads $\{F^e\}$ thus relating $\{F^e\}$ to $\{\delta^e\}$

$$\{F^e\} = \int [B]^T[D][B]\, d(\text{vol})\, \{\delta^e\} \tag{VI}$$

By comparison with equation I we see that

$$[K^e] = \int [B]^T[D][B]\, d(\text{vol})$$

Step VII

Establish a stress–displacement matrix $[H]$

$$\{\sigma(x, y)\} = [H]\{\delta^e\} \tag{VII}$$

where, from Step V,

$$[H] = [D][B]$$

Matrix Algebra

This appendix provides a very brief introduction to matrix algebra, dealing only with the points encountered in this text. For readers unfamiliar with matrix methods a list of texts providing a fuller presentation is given at the end of this appendix.

Definition

A matrix is an array of terms as shown in equation A.1. The terms may be either pure numbers, constants or variables, or a mixture of all three as shown in equation A.2.

$$[A] = \begin{bmatrix} a_{11} & a_{12} & \cdot & \cdot & \cdot & a_{1n} \\ a_{21} & a_{22} & \cdot & \cdot & \cdot & a_{2n} \\ \cdot & \cdot & \cdot & \cdot & \cdot & \cdot \\ \cdot & \cdot & a_{ij} & \cdot & \cdot & \cdot \\ a_{m1} & a_{m2} & \cdot & \cdot & \cdot & a_{mn} \end{bmatrix} \tag{A.1}$$

$$[B] = \begin{bmatrix} 3 & a^2 & \dfrac{x}{2} \\ y^2 & x+4 & b \\ x+y^2 & 2 & 1+y \end{bmatrix} \tag{A.2}$$

The matrix of equation A.1 possesses m rows and n columns, and is therefore said to be of order $m \times n$. The term appearing in the ith row and the jth column is denoted by a_{ij}. When $m = n$ the matrix is termed a square matrix. Such matrices possess special properties and are of particular significance in structural theory. If $n = 1$, the matrix reduces to a single column and is sometimes termed a 'column vector', and column vector $\{B\}$ in equation A.3 is a typical example. When $m = 1$, we have a row vector as shown in equation A.4.

$$\text{Column vector } \{B\} = \begin{Bmatrix} 2 \\ 3x \\ a+4 \\ y^3 \end{Bmatrix} \tag{A.3}$$

$$\text{Row vector } \{A\} = \{x \quad 6+y \quad 3y^2\} \tag{A.4}$$

The terms a_{ii} lie on the leading diagonal and are called the main diagonal terms. In equation A.5, these consist of the terms $1, x+4, 3$.

$$[C] = \begin{bmatrix} 1 & 2x^2 & 3 \\ 4x & x+4 & 0 \\ -6 & -9x & 3 \end{bmatrix} \tag{A.5}$$

Transpose

The transpose of a matrix $[A]$ is denoted by $[A]^T$ and is obtained by interchanging rows and columns. Therefore, if we take equation (A.1) as an example, $[A]^T$ becomes

$$[A]^T = \begin{bmatrix} a_{11} & a_{21} & \cdot & \cdot & \cdot & a_{m1} \\ a_{12} & a_{22} & \cdot & \cdot & \cdot & a_{m2} \\ \cdot & \cdot & \cdot & \cdot & \cdot & \cdot \\ \cdot & \cdot & \cdot & \cdot & \cdot & \cdot \\ \cdot & \cdot & \cdot & \cdot & \cdot & \cdot \\ a_{1n} & a_{2n} & \cdot & \cdot & \cdot & a_{mn} \end{bmatrix} \tag{A.6}$$

In particular, the transpose of a column vector becomes a row vector.

Addition and subtraction

Matrices can only be added or subtracted if they are of the same order. Thus an $m \times n$ matrix may not be added to a $p \times q$ matrix. The process is simply to add the corresponding terms. Therefore, if $[C]$ is the sum of the matrices $[A]$ and $[B]$, $[C]$ will be given by

$$c_{ij} = a_{ij} + b_{ij}$$

Thus

$$\begin{bmatrix} a_{11} & a_{12} \\ a_{21} & a_{22} \end{bmatrix} + \begin{bmatrix} b_{11} & b_{12} \\ b_{21} & b_{22} \end{bmatrix} = \begin{bmatrix} (a_{11}+b_{11}) & (a_{12}+b_{12}) \\ (a_{21}+b_{21}) & (a_{22}+b_{22}) \end{bmatrix}$$

hence

$$c_{11} = a_{11}+b_{11} \qquad c_{12} = a_{12}+b_{12}$$
$$c_{21} = a_{21}+b_{21} \qquad c_{22} = a_{22}+b_{22}$$

Example 1

$$\text{Matrix } [A] = \begin{bmatrix} 5 & -4 & 7 \\ -8 & 5 & 6 \end{bmatrix}$$

is to be added to

$$\text{matrix } [B] = \begin{bmatrix} 1 & 3 & -6 \\ 4 & 5 & 8 \end{bmatrix}$$

$$[A]+[B] = [C]$$

$$\begin{bmatrix} 5 & -4 & 7 \\ -8 & 5 & 6 \end{bmatrix} + \begin{bmatrix} 1 & 3 & -6 \\ 4 & 5 & 8 \end{bmatrix} = \begin{bmatrix} 6 & -1 & 1 \\ -4 & 10 & 14 \end{bmatrix}$$

Example 2

Matrix $[B]$ is to be subtracted from matrix $[A]$ to give a matrix $[C]$, i.e. $[C] = [A]-[B]$.

In this case a term such as c_{ij} has the value

$$c_{ij} = a_{ij} - b_{ij}$$

Thus

$$\begin{bmatrix} 5 & -4 & 7 \\ -8 & 5 & 6 \end{bmatrix} - \begin{bmatrix} 1 & 3 & -6 \\ 4 & 5 & 8 \end{bmatrix} = \begin{bmatrix} 4 & -7 & 13 \\ -12 & 0 & -2 \end{bmatrix}$$

Example 3

$2[D] + 5[E] - 3[F]$

$$= 2\begin{bmatrix} 1 & 2 \\ 3 & 4 \end{bmatrix} + 5\begin{bmatrix} -5 & 6 \\ -7 & 8 \end{bmatrix} - 3\begin{bmatrix} -9 & 8 \\ 7 & -6 \end{bmatrix}$$

$$= \begin{bmatrix} 2 & 4 \\ 6 & 8 \end{bmatrix} + \begin{bmatrix} -25 & 30 \\ -35 & 40 \end{bmatrix} + \begin{bmatrix} 27 & -24 \\ -21 & 18 \end{bmatrix}$$

$$= \begin{bmatrix} 4 & 10 \\ -50 & 66 \end{bmatrix}$$

In certain circumstances where two matrices of different order must be added, it is permissible to increase the size of the smaller by inserting rows and columns of zeros to make both matrices the same size. Examples of this are given in Chapter 2.

Multiplication

Two matrices may be multiplied together only if the number of columns in the first is equal to the number of rows in the second: e.g. an $m \times n$ matrix may be multiplied by an $n \times p$ matrix to give an $m \times p$ matrix. The terms in the product matrix resulting from the multiplication of a matrix $[A]$ with a matrix $[B]$ are obtained by taking the scalar product of each row of matrix $[A]$ with each column of matrix $[B]$:

$$c_{ij} = \sum_{k=1}^{n} a_{ik} b_{kj} \tag{A.7}$$

Thus

$$\begin{bmatrix} a_{11} & a_{12} \\ a_{21} & a_{22} \end{bmatrix} \begin{bmatrix} b_{11} & b_{12} \\ b_{21} & b_{22} \end{bmatrix} = \begin{bmatrix} (a_{11}b_{11}+a_{12}b_{21}) & (a_{11}b_{12}+a_{12}b_{22}) \\ (a_{21}b_{11}+a_{22}b_{21}) & (a_{21}b_{12}+a_{22}b_{22}) \end{bmatrix}$$

Hence

$$c_{11} = a_{11}b_{11}+a_{12}b_{21} \qquad c_{12} = a_{11}b_{12}+a_{12}b_{22}$$
$$c_{21} = a_{21}b_{11}+a_{22}b_{21} \qquad c_{22} = a_{21}b_{12}+a_{22}b_{22}$$

Let us consider the following example.

$$[A] = \begin{bmatrix} 1 & 2 \\ 4 & 7 \end{bmatrix} \qquad [B] = \begin{bmatrix} 1 & 3 & -6 \\ 4 & 5 & 8 \end{bmatrix}$$

$$\therefore [A][B] = \begin{bmatrix} 1 & 2 \\ 4 & 7 \end{bmatrix} \begin{bmatrix} 1 & 3 & -6 \\ 4 & 5 & 8 \end{bmatrix}$$

Here $m=2$, $n=2$ and $p=3$, and $[C]$ will therefore be a 2×3 matrix. To obtain the first term of $[C]$, set $i=j=1$ in equation (A.7).

$$c_{11} = a_{11}b_{11}+a_{12}b_{21} = 1\times1+2\times4 = 9$$

Thus we see that the term c_{11} is obtained by considering the matrix product of equation (A.8).

$$c_{11} = \begin{bmatrix} 1 & 2 \end{bmatrix} \begin{bmatrix} 1 \\ 4 \end{bmatrix} \qquad\qquad (A.8)$$

We need therefore consider only the row and column concerned in the multiplication. If this procedure is continued we obtain $[C]$ as

$$C = \begin{bmatrix} 9 & 13 & 10 \\ 32 & 47 & 32 \end{bmatrix}$$

The minus sign in matrix $[B]$ presents no special difficulty, it is simply necessary to take note of the signs of the terms when performing the summation.

An aid to those who experience difficulty in matrix multiplication is given below:

$$
\begin{array}{cc}
 & \text{matrix } [B] \\
\begin{bmatrix} 1 & 3 & -6 \\ 4 & 5 & 8 \end{bmatrix} & \swarrow \\
\end{array}
$$

matrix $[A]$
$$
\begin{bmatrix} 1 & 2 \\ 4 & 7 \end{bmatrix}
\qquad
\begin{bmatrix} c_{11} & c_{12} & c_{13} \\ c_{21} & c_{22} & c_{23} \end{bmatrix}
$$
product matrix $[C]$

To obtain c_{22}

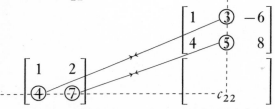

c_{22} is given by the summation of the scalar product of row 2 with column 2, as indicated by the arrows: i.e. $c_{22}=4\times3+7\times5=47$. Likewise, $c_{13}=1\times-6+2\times8=10$.

Transpose of a product

It is often necessary to take the transpose of a matrix product. It can be established that the transpose of a matrix product of matrix $[A]$ and matrix $[B]$ is given by the product of $[B]^T$ and $[A]^T$: i.e. $([A][B])^T=[B]^T[A]^T$. (Attention is drawn to the reversal of the order of the matrices $[A]$ and $[B]$.) This can be readily shown to be correct by the following example.

The product of

$$
\text{matrix } [A] = \begin{bmatrix} 1 & 2 \\ 3 & 4 \end{bmatrix}
\quad \text{and of} \quad
\text{matrix } [B] = \begin{bmatrix} 5 & 6 \\ 7 & 8 \end{bmatrix}
$$

$$[A][B] = [C]$$

$$
\begin{bmatrix} 1 & 2 \\ 3 & 4 \end{bmatrix}
\begin{bmatrix} 5 & 6 \\ 7 & 8 \end{bmatrix}
=
\begin{bmatrix} 19 & 22 \\ 43 & 50 \end{bmatrix}
$$

Let us now obtain the product of $[B]^T[A]^T=[D]$.

$$\begin{bmatrix} 5 & 7 \\ 6 & 8 \end{bmatrix}\begin{bmatrix} 1 & 3 \\ 2 & 4 \end{bmatrix} = \begin{bmatrix} 19 & 43 \\ 22 & 50 \end{bmatrix}$$

Thus we see that $[D]=[C]^T$.

Symmetric matrix

A very important type of square matrix, of particular interest in connection with structural problems, is a symmetric matrix. Such matrices possess the property

$$a_{ji} = a_{ij}$$

They can therefore be completely specified by the terms on the main diagonal, i.e. those of the form a_{ii}, together with those above the main diagonal, i.e. those for which $j>i$.

Matrices and simultaneous equations

Undoubtedly the most attractive feature of matrix algebra when used for purposes of structural analysis is the way it enables the engineer to write down large systems of simultaneous equations in a simple and straightforward manner. This permits standard computer programs to be written which are sufficiently general to enable one program to analyse several different types of structures.

Consider the set of equations:

$$x + 2y + z = 8$$
$$2x - y + 3z = 7$$
$$x + y - z = 4$$

Remembering the rule for matrix multiplication stated earlier, we may rewrite these equations in matrix form as:

$$\begin{bmatrix} 1 & 2 & 1 \\ 2 & -1 & 3 \\ 1 & 1 & -1 \end{bmatrix}\begin{Bmatrix} x \\ y \\ z \end{Bmatrix} = \begin{Bmatrix} 8 \\ 7 \\ 4 \end{Bmatrix}$$

or in compact form as

$$[A]\{\alpha\} = \{B\} \tag{A.9}$$

where $[A]$, $\{\alpha\}$ and $\{B\}$ refer to the square matrix and the two-column vectors, respectively. The solution of the set of equations then corresponds to

$$\{\alpha\} = [A]^{-1}\{B\} \tag{A.10}$$

In equation A.10, $[A]^{-1}$ corresponds to the inverse of $[A]$. The inversion of a matrix is an analogous process to the division of a scalar quantity. The inverse matrix is defined by

$$[A]^{-1}[A] = [I]$$

where $[I]$ is the identity matrix, so called because $[B][I]=[B]$. The identity matrix consists of a series of unit values down the leading diagonal with zeros elsewhere, as shown below for a 3×3 matrix.

$$[I] = \begin{bmatrix} 1 & 0 & 0 \\ 0 & 1 & 0 \\ 0 & 0 & 1 \end{bmatrix}$$

Using this definition we may obtain equation A.10 from equation A.9 simply by premultiplying each side by $[A]^{-1}$ to give

$$[A]^{-1}[A]\{\alpha\} = [A]^{-1}\{B\}$$
$$[I]\{\alpha\} = [A]^{-1}\{B\}$$

It now remains to discuss the quantity $[A]^{-1}$ in detail.

Inverse of a matrix

Only square matrices possess an inverse. This inverse is itself square and of the same size as the original matrix. The inverse only exists if the matrix is non-singular, i.e. if its determinant is non-zero. The inversion of matrices and the parallel problem of solving sets of linear equations are important topics in numerical analysis, and many different methods for performing these tasks exist. The interested reader is referred to the text by Faddeeva[1] for details of several of these methods. In the field of structural analysis the methods of Gaussian elimination, Gauss–Seidel iteration and Cholesky square-root are among those that have been used successfully.

Matrix differentiation

Often in this text the derivative of a matrix is required. The rule for this is very simple. A matrix may be differentiated simply by differentiating each term in the normal manner. Therefore, if $[A]$ is given by

$$[A] = \begin{bmatrix} x^2 & 3x^2 & 4x \\ 6 & \frac{1}{2}x^2 & 5x \\ 2x^3 & x^4 & 2 \end{bmatrix}$$

$$\frac{d}{dx}[A] = \begin{bmatrix} 2x & 6x & 4 \\ 0 & x & 5 \\ 6x^2 & 4x^3 & 0 \end{bmatrix}$$

Matrix integration

In the same way, a matrix may be integrated simply by integrating each term. Using the matrix $[A]$ defined previously:

$$\int_0^x [A]\, dx = \begin{bmatrix} \dfrac{x^3}{3} & x^3 & 2x^2 \\ 6x & \dfrac{x^3}{6} & \dfrac{5x^2}{2} \\ \dfrac{x^4}{2} & \dfrac{x^5}{5} & 2x \end{bmatrix}$$

More complex matrix expressions of the form shown below must be integrated by first performing the multiplication and then integrating each term of the resulting matrix as indicated above. This process is essentially similar to that for integrating the product of a series of algebraic expressions:

$$\int_0^x \int_0^y [B]^T [D][B]\, dx\, dy$$

e.g. $\int_0^x [B]^T [A] \, dx$ where

$$[B] = \begin{bmatrix} 2x & x^2 \\ 4 & 2x \end{bmatrix} \qquad [A] = \begin{bmatrix} 3x & 2 \\ x^2 & 1 \end{bmatrix}$$

$$\int_0^x [B]^T [A] \, dx = \int_0^x \begin{bmatrix} 2x & 4 \\ x^2 & 2x \end{bmatrix} \begin{bmatrix} 3x & 2 \\ x^2 & 1 \end{bmatrix} dx$$

$$= \int_0^x \begin{bmatrix} 10x^2 & 4x+4 \\ 5x^3 & 2x^2+2x \end{bmatrix} dx$$

$$= \begin{bmatrix} \dfrac{10x^3}{3} & 2x^2+4x \\[2ex] \dfrac{5x^4}{4} & \dfrac{2x^3}{3}+x^2 \end{bmatrix}$$

When a double integral sign occurs, the integration must follow normal calculus procedures.

References

1. FADDEEVA, V. N. *Computational Methods of Linear Algebra*. New York, Dover Publications Co., 1959.
2. GERE, J. M. and WEAVER, W. *Matrix Algebra for Engineers*. Princetown, N.J., D. Van Nostrand Co. Inc., 1965.
3. FROBERG, C. E. *Introduction to Numerical Analysis*. Reading, Mass., Addison-Wesley Publishing Co. Inc., 1965.
4. LEE, J. A. N. *Numerical Analysis for Computers*. New York, Reinhold Publishing Corp., 1966.

Index